Infrared and
Raman Spectroscopy

PRACTICAL SPECTROSCOPY

A SERIES

Edited by Edward G. Brame, Jr.

Elastomer Chemicals Department
Experimental Station
E.I. du Pont de Nemours and Co., Inc.
Wilmington, Delaware

Volume I: Infrared and Raman Spectroscopy (in three parts)
edited by Edward G. Brame, Jr. and Jeanette G. Grasselli

ADDITIONAL VOLUMES IN PREPARATION

Infrared and Raman Spectroscopy

(in three parts)

PART B

Edited by

EDWARD G. BRAME, Jr.

Elastomer Chemicals Department
Experimental Station
E. I. du Pont de Nemours and Co., Inc.
Wilmington, Delaware

JEANETTE G. GRASSELLI

The Standard Oil Company
Research and Engineering Department
Cleveland, Ohio

MARCEL DEKKER, INC.   New York and Basel

Library of Congress Cataloging in Publication Data
Main entry under title:

Infrared and Raman spectroscopy.

  (Practical spectroscopy ; v. 1)
  Includes bibliographies.
  1. Infra-red spectrometry. 2. Raman spectroscopy.
I. Brame, Edward G., 1927-   II. Grasselli,
Jeanette G. III. Series. [DNLM: 1. Spectrophoto-
metry, Infrared. 2. Spectrum analysis, Raman.
QC457 I43]
QD96.I5I53   535'.842   75-32391
ISBN 0-8247-6526-5

27,655

Pt. B

MARCEL DEKKER, INC.

270 Madison Avenue, New York, New York 10016

Current printing (last digit):
10 9 8 7 6 5 4 3 2 1

PRINTED IN THE UNITED STATES OF AMERICA

PREFACE

This book is the second one of a new series on spectroscopy. It is Part B of Volume I and like Part A is devoted to the fields of infrared and Raman spectroscopy. Part C of Volume I will be published subsequently.

The aim and objective of this new series is to cover all the various applications of spectroscopy and to illustrate its usefulness in such important and diverse areas as organic chemistry, inorganic chemistry, polymer chemistry, biological science, environmental science, food, textiles, etc. Through the discussions in various chapters, the reader will gain a better appreciation and understanding of the different uses to which spectroscopy has been applied. Also, it is hoped that this series will not only provide a practical reference to applications of spectroscopy but will also encourage a cross fertilization between applications in the different fields of biology, chemistry, and physics.

Volume I, Part B contains six chapters. The first one, authored by Dr. R. P. Young, discusses computer systems. It describes data acquisition from off-line and in-line spectrometers as well as the different methods of handling the acquired data for both qualitative and quantitative analysis.

Chapter 2 is coauthored by Drs. R. Nyquist and R. O. Kagel and deals with organic materials. It covers a large number of applications and demonstrates the highly successful role that infrared and Raman spectroscopy continue to play in the analysis of these materials.

Chapter 3, authored by Dr. D. S. Lavery, covers the important uses of infrared and Raman spectroscopy in environmental science. Topics discussed include sample handling, calibration methods, remote sensing, and multicomponent analyses.

Chapter 4, authored by Dr. A. Eskamani, discusses the use of infrared and Raman spectroscopy in the food industry. It deals with the analysis of food materials and describes the techniques that are being used and the results that can be obtained from the analyses.

Chapter 5 on the subject of petroleum is authored by Mr. P. B. Tooke. This chapter describes the many and varied applications of infrared and Raman spectroscopy in the analysis of petroleum products.

Chapter 6, authored by Mr. G. Celikiz, covers the use of infrared and Raman spectroscopy in textiles. It may be a surprise to many spectroscopists that infrared especially has been used very successfully in the textile industry. This chapter reports that use.

Since the current volume initiates a brand new series called *Practical Spectroscopy,* it serves the dual purpose of showing the format and approximate subject coverage for future books. Subsequent volumes will cover applications of x-ray spectroscopy, mass spectroscopy, nuclear magnetic resonance spectroscopy, ultraviolet-visible spectroscopy, emission spectroscopy, etc., in the various fields of chemistry. These volumes will be published as the material is collected, edited, and prepared for publication on an ongoing basis. There will not be a rigid schedule for publication, and some of the books may be published simultaneously.

One final objective is to bring to the practicing spectroscopist a consistent and practical approach in illustrating the many and varied uses of spectroscopy. In turn, the chemist, professor, or graduate student will find that this series provides a useful overview on the state of the art in applications of spectroscopic methods.

E. G. Brame, Jr.

CONTRIBUTORS TO PART B

Gultekin Celikiz, Department of Chemistry, Philadelphia College of Textiles and Science, Philadelphia, Pennsylvania

A. Eskamani, Analytical Division, Research and Development Center, Standard Oil Company (Ohio), Cleveland, Ohio

R. O. Kagel,[*] Analytical Laboratories, Dow Chemical Company, Midland, Michigan

Donald S. Lavery,[†] Marketing Department, Wilks Scientific Corporation, South Norwalk, Connecticut

Richard A. Nyquist, Analytical Laboratories, Dow Chemical Company, Midland, Michigan

P. B. Tooke, Research Center, British Petroleum Company, Sunbury-on-Thames, Middlesex, England

R. Peter Young,[‡] Perkin-Elmer, Ltd., Beaconsfield, England

[*]Present address: Environmental Services, Dow Chemical Company, Midland, Michigan

[†]Present affiliation: Consultant in Analytical Chemistry, East Norwalk, Connecticut

[‡]Present address: ICI Corporate Laboratory, Runcorn, England

CONTENTS

Preface iii

Contributors to Part B. v

Contents of Other Parts ix

Chapter 5: COMPUTER SYSTEMS. 347

              R. Peter Young

        I. Introduction. 348
       II. Digital Data Acquisition. 350
      III. Data Refinement 378
       IV. Methods of Quantitative Analysis. 399
        V. Band Shape Analysis 417
       VI. Spectrum Storage and Retrieval. 425
            References. 434

Chapter 6: ORGANIC MATERIALS 441

              Richard A. Nyquist and R. O. Kagel

        I. Introduction. 442
       II. Experimental. 443
      III. Group Frequencies in Molecular Structure
            Determinations. 463
            References. 558
            Appendix: Indexes of Spectra and Charts. . . . 560

Chapter 7: ENVIRONMENTAL SCIENCE 565

              Donald S. Lavery

        I. Introduction. 565
       II. Gas Analysis. 567
      III. Water Analysis. 615
            References. 621

Chapter 8: FOOD INDUSTRY 623

              A. Eskamani

        I. Introduction. 623
       II. Determination of Water in Food. 624
      III. Analysis of Food Carbohydrates. 628

    IV. Analysis of Food Proteins. 635
     V. Applications to Food Lipids. 638
    VI. Identification of Food Flavors 651
   VII. Analysis of Vitamins 657
  VIII. Analysis of Additives. 659
    IX. Conclusion 661
         References 662

Chapter 9: PETROLEUM. 667

            P. B. Tooke

     I. Introduction 667
    II. Scope of Infrared Applications 669
   III. Fuels. 670
    IV. Lubricants 672
     V. Raman Spectroscopy 697
    VI. Future Developments. 699
         References 700

Chapter 10: TEXTILES 701

            Gultekin Celikiz

     I. Introduction 702
    II. Methods for Obtaining Infrared Spectra 702
   III. Applications of Infrared Spectroscopy. 709
    IV. Conclusion 712
         References 713

Cumulative Index appears in Part C

CONTENTS OF OTHER PARTS

Part A

An Introduction to Molecular Vibrations, Bryce Crawford, Jr. and Douglas Swanson

Inorganic Materials, Robert L. Carter

Organometallic Compounds: Vibrational Analysis, Walter F. Edgell

Ionic Organometallic Solutions, Walter F. Edgell

Part C

Biological Science, George J. Thomas, Jr. and Yashimasa Kyogoku

Polymers, Sandra C. Brown and Albert B. Harvey

Surfaces, Clara D. Craver

Infrared and
Raman Spectroscopy

Chapter 5

COMPUTER SYSTEMS

R. Peter Young*

Perkin-Elmer, Ltd.
Beaconsfield, England

I.   INTRODUCTION 348

II.   DIGITAL DATA ACQUISITION 350
      A. Basic Hardware and Terminology 350
      B. Sampling Considerations. 356
      C. Off-line Data Logging. 362
      D. On-line Spectrometer Systems 369

III. DATA REFINEMENT. 378
      A. Calibration and Formatting 378
      B. Smoothing and Sharpening 381
      C. Determination of Optical Constants 391

IV.   METHODS OF QUANTITATIVE ANALYSIS 399
      A. Heights and Areas. 399
      B. Difference and Derivative Spectra. 401
      C. Multicomponent Mixtures. 410

V.   BAND SHAPE ANALYSIS. 417
      A. Method of Truncated Moments. 417
      B. Band Fitting (II). 419
      C. Dipole Correlation Functions 421

VI.   SPECTRUM STORAGE AND RETRIEVAL 425
      A. Archival Records 425
      B. Spectrum Condensates 426
      C. Library Search Methods 430

      REFERENCES 434

*Present address: ICI Corporate Laboratory, Runcorn, England.

I. INTRODUCTION

Spectroscopists realized at a very early date that the digital electronic computer was a tool so powerful it could hardly fail to rev-
olutionize all the data-handling aspects of their trade. These
initial expectations have not been fully realized even yet, since
computer systems for infrared (IR) and Raman spectroscopy have
evolved only slowly. The type of computing machine which evolved
in the mid- to late sixties, characterized by a fairly massive hardware configuration and cumbersome input-output facilities, was not
particularly compatible with spectroscopic needs. The development
of high-level scientific languages such as Fortran and Algol allowed
very sophisticated data analysis programs to be written even for
these machines and represented a great stride forward. Even so, the
software requirements for IR and Raman spectroscopy were by no means
as clear-cut as for gas chromatography (GC), for example. In this
principally academic environment, there was little commonality, except at a rather superficial level, between the programs written and
used by the various spectroscopists who were actively using computers.

Consequently, even when the advent of the minicomputer opened
the way to dramatic new possibilities in laboratory automation, IR
spectroscopy lagged behind GC, where an easily definable software
package could almost immediately be implemented, and certain other
areas [high-resolution mass spectroscopy (MS) and Fourier transform
(FT) techniques] where the computer is essential for the high-speed
acquisition and preprocessing of raw data. There are still very few
broadly applicable, general purpose IR data systems combining both
hardware and software into a single package. Certain of the interferometer systems approach this target but their present high cost
seems likely to deter those applications not requiring the Fourier
mode of operation.

Thus, the present situation is that IR and Raman spectroscopists
who wish to computerize their instrumentation generally put together
systems for their own particular requirements, often using computer

facilities already available, and by writing their own special
purpose programs. This has inevitably led to a diversity of custom-
built spectrometer/computer hardware configurations with, perhaps,
an equal number of software variants even allowing for the inter-
system compatibility of programs written in the standard languages.
We will try here, therefore, to discuss the relative merits and difficulties associated with identifiable types of hardware systems
rather than be concerned with particular instruments or machines.
Similarly, we will describe some of the methods used for data acquisition and instrument interfacing without discussing the details
which are better obtained from manufacturers' manuals.

As for software, we shall deal with algorithms rather than
actual programs, since the translation between one machine or lan-
guage and another is more mechanical than conceptual. Moreover,
the original algorithms are invariably consulted when programming
at the assembler level. In many cases, an algebraic or verbal description will suffice to clarify the individual steps of a particular computational procedure.

From the spectroscopic standpoint, we shall touch upon the problems associated with the physical means of measurement (transmission,
reflection, attenuated total reflection (ATR), or inteferometry),
since computer correction or preprocessing of the experimentally
observed data can have a valid part to play in each technique. Because of its greater general applicability, we will treat in more
detail the various means of data analysis and interpretation. These
will include spectrum addition and subtraction, band fitting, multicomponent mixture analysis, and band shape measurements, together
with noise smoothing and resolution enhancement.

Finally, we shall deal with possible means of spectrum storage
and retrieval, particularly as applied to compound identification
by searching libraries of coded spectra for exact or close resem-
blances to an unknown sample.

II. DIGITAL DATA ACQUISITION

A. Basic Hardware and Terminology

1. *The Computer and its Peripherals*

The internal structure and logical organization of digital
computers (i.e., their architecture) is continually evolving, each
generation being distinguished from the next by the major advances
in solid state technology. Two functional elements common to all
are the central processor unit (CPU) and the main memory, which are
linked together by an intricate system of data and control lines.
Programs held in the memory consist of sequences of instructions in
the form of binary words, which are decoded at the time of execution
by the CPU, in whose registers the required arithmetic or logic operations are performed on data obtained from elsewhere in the memory.
Each memory location is associated with a numeric address, which
may be specified directly or indirectly in the instruction word.

The efficiency of the machine is largely determined by its
word length, e.g., 12, 16, or 24 bits (which affects the complexity
of operation that can be expressed in a single instruction, the
number of memory locations that are directly addressable, and the
range or precision of the numbers that can be handled as data), and
by the speed with which a given operation can be performed. This
is often quoted in terms of memory cycle time, typically in the
range 3 microseconds (μs) to 300 nanoseconds (ns). Overall capability is reckoned by the capacity of main memory, i.e., core or semi-
conductor, in terms of the number of words (4K equals 4,096, 8K
equals 8, 192, and so on) because this determines the length of programs that can be run and the amount of immediately addressable data
that can be stored; but the core size is generally extendable to 16K
or 32K, with this option meeting practically all spectroscopic requirements.

Such a system is useful to us only if it can take in data
generated by the spectrometer and output the computed results in
some suitable form. It is therefore necessary to provide a range
of peripheral devices, such as a tape reader, character printer,

oscilloscope, and operator console, each of which communicates with
the CPU through a special hardware interface. The ubiquitous tele-
typewriter (TTY) functionally serves as paper tape reader and punch,
plus keyboard and printer, but also features a disturbing noise
level and a low data transmission rate of only 10 characters/sec.
Faster, virtually silent keyboard/printers (at 30 or more characters/
sec) are available at higher cost, as also are visual display terminals with character-only or full graphics capabilities. The types
and relative merits of punched paper and magnetic tape systems are
discussed in more detail later (Sec. II.C.3, 4).

Perhaps the most important computer peripheral device for the
spectroscopist is the disk or tape backing store which can be regarded as a large memory extension that is not quite instantaneously
addressable. This is typically a magnetic disk capable of holding
up to a million words of programs or data, formatted in many con-
centric tracks on the disk surface. Data are transferred from the
spinning disk through a single read/write head that can be rapidly
driven to any track position (moving head-type) or, alternatively,
by any of a bank of stationary read/write heads (fixed head-, or
head-per-track-type). The latter are more costly, but because the
head movement time is eliminated, they allow much faster access to
any given block of data. For either type of so-called random access
memory the data seek time is so low [e.g., 20 or 50 milliseconds
(ms)] and the data transfer rate so high (e.g., over 100K characters/
sec), that the alternative sequential access magnetic tape systems
seem pedestrian by comparison. It should always be borne in mind,
however, that for use simply as a data bank or spectrum library, a
delay of even 30 sec may be quite acceptable, e.g., to retrieve a
previously stored spectrum. This is especially relevant in view of
the current low cost of, for example, magnetic tape cassette systems.

2. Codes and Communications

We are accustomed to dealing with spectral information in analog
form as a continuous graphical representation of intensity as a
function of wavelength or wave number. In order to present this
information to a computer, it must first be sampled at discrete

abscissa intervals (Sec. II.B.1) and converted into digital form by representing the data in a suitably formatted binary code. When these digital signals are fed directly to the CPU via an interface, i.e., the spectrometer is on-line to the computer (Sec. II.D), it is usual to transmit the data as binary numbers, each of sufficient length to code the maximum value likely to be encountered to the required precision (Sec. II.B.2).

One disadvantage of using pure binary code is that consecutive numbers generally differ in several bit positions. Therefore, if the data source is fluctuating slightly and there is an error in one of these bits at the instant of sampling, the resulting number may be very different from either approximate value. To lessen this risk, so-called Gray codes have been devised in which all consecutive pairs of numbers differ by one bit only. These are frequently used for the output of shaft encoders, which are commonly employed in the digital recording of wave number position. Because this changes smoothly during a spectrum scan, a useful degree of error checking can be incorporated into the interface.

It is often reassuring to monitor visually the data that are being sampled and transmitted through a digital readout or numeric character display (e.g., Nixie- or LED-type). To drive such a display, the data must be converted at some stage to a binary coded decimal (BCD) format, in which four bits are allocated to each digit of the decimal value, with special provision for decimal point and plus and minus characters.

We have so far only considered the coding of numeric data. Just as important is the ability to transmit alphabetic and other characters used in names, program instructions, and so on. The full alphanumeric set of over 60 printable and control characters may be handled by the American Standard Code for Information Interchange (ASCII), which is the code used by all TTYs and most time-sharing systems. It should be noted that some manufacturers have developed different codes for their own machinery (e.g., IBM uses EBCDIC code). The ASCII code uses seven bits per character, which plus an optional

one for parity checking makes an eight-bit unit commonly called a
byte. The ASCII has become a standard code for off-line logging of
laboratory data (Sec. II.C). Paper tapes punched in ASCII code are
directly listable on a TTY and can, moreover, easily be checked by
visual inspection. With the sprocket hole to the right, the right-
most four binary positions directly give the value of a decimal numeric character.

The TTY compatability of the ASCII code enables such data to
be transmitted by an appropriate terminal over private land lines
to a remote (or central) computer facility, through the national
telephone system to any commercial time-sharing service, or even
literally around the world using international satellite links (for
example, the Honeywell Mark III Network service has interconnected
centers in Europe, America, and Japan). Indeed, a data logger and
time-sharing terminal probably represent the minimum investment configuration for those willing to write their own programs and who
have relatively little data to transmit (see also Sec. II.C.1).

   3. Multiplex and Front-end

When several instruments or devices operate apparently simultaneously in conjunction with a single central processor, they are
said to be multiplexed. As soon as incoming data are ready to be ac-
cepted by the computer, it takes only a matter of microseconds to
complete the actual transter into one of the registers of the CPU.
Even with sampling rates of 10 readings/sec, which is fast (spectro-
scopically) for all except FT operation, the CPU is busy with a given
instrument for only 0.01% of the time. Consequently there is ample
time to interleave the input requirements of a large number of such
instruments if the need arises. In practice, it is more usual for
instruments of quite different data rates to be multiplexed together,
and a proper system of priority interrupt procedures must be used to
ensure satisfactory operation of the whole system (Sec. II.D.5).

Just as several spectrometers may be multiplexed to a single
central processor, so too may several small computers be multiplexed
to a much larger one. This situation can often arise when an estab-

lishment having a large computer center wishes to decentralize the
relatively trivial or repetitive tasks of data acquisition and refinement, and only accept for itself the complex analysis or data
banking requirements. The intermediate machine is then described as
a "front-end" to the major processor. In the general time-sharing
environment, a laboratory may have only a single computer, dedicated
to perhaps several instruments but having access to a larger system,
in which case it might alternatively be described as an intelligent
terminal. It is then the job of the time-sharing supervisor program
(or executive) to control the multiplexing of the various intelligent
terminals that may be connected at any one time.

The term "front-end" is also used in connection with interface
design when it refers to how the CPU looks to an external device.
It is an electrical and logical specification of that part of the
computer being brought outside to meet the device at the interface.

4. Minicomputer Systems

The introduction in the late sixties and the subsequent rapid
and highly successful development of the minicomputer has been
watched or enjoyed by many of those interested in laboratory automation in general and spectroscopy in particular. Here, suddenly,
was a bench-top machine which could do even faster all that its
predecessor, many times larger, could do, and at an equivalent fraction of the cost. This encouraged a shift away from the early,
rather cumbersome, multiplexed configurations towards dedicated systems where each major instrument could almost afford to have its own
computer right alongside it in the laboratory. It was perhaps this
achievement of local autonomy, i.e., full control of the overall
system firmly in the hands of the laboratory staff, which was felt to
be the most significant new factor.

At the present time, practically all the minicomputers on the
market have a long line of peripherals in support, with interfaces
available off-the-shelf, including paper tape reader/punch units,
large and small magnetic tape systems, fixed head, moving head, and
cartridge disc drives, plus keyboard/printers and visual display

terminals for operator communications. Although a highly sophisticated system of a minicomputer and its peripherals is perhaps the
only realistic system to consider (see Sec. II.D.4), it should be
noted that the total cost of the computing hardware is likely to
exceed that of the spectrometer itself. Moreover, the applications
software will generally involve additional expense or effort.

This brings us partway back to the previous situation where,
to keep overall costs down, the system must be shared by several
instruments. IR and Raman spectroscopy, having a less urgent requirement than some other techniques, tend to appear at the lower
end of the priority interrupt list (Sec. II.D.5). It will be interesting to see how long this situation will continue when the next
generation of even cheaper computers and peripherals finally emerges.

5. *The Spectrometer Interface*

We conclude the present discussion with some remarks about the
spectrometer side of the hardware interface, but leave to the next
section (II.B) a fuller treatment of the spectroscopic implications
of how the data are sampled.

We consider first the requirements of the interface to an off-
line data logging unit. Its principal functions are to sample the
intensity value at a preselected rate (i.e., at fixed time or wave
number intervals), together with the actual wave number position
(although this, as we shall see in Sec. II.B.3, should be regarded
as optional), convert each reading into a string of coded characters
(e.g., ASCII), and arrange them according to a specified format. In
double beam IR instruments based on the optical-null principle, the
intensity value (usually transmittance) may be taken by either a
linear or a shaft encoder, unless an external readout point is provided, when it is better to go directly to a digital voltmeter (DVM)
or simple A/D converter. For reading the wave number position, a
shaft encoder is normally used and fitted to the prism or grating
drive mechanism. If ordinate data only are sampled on a fixed time
interval basis, it is important not only to ensure that the sample
timer and scan drive are in proper synchronism, but also to inhibit

digitization during any pauses, such as commonly occur during filter and grating changeovers.

If on the other hand we wish to interface the spectrometer directly to the computer and include some degree of instrument control, the requirements are somewhat different and the opportunities greater. The sampling of the spectrum coordinates must still be done. With a conventional wave number drive fitted with an encoder, this could be organized on an interrupt basis, or if a computer-controlled stepper motor is used, this could be triggered from its drive pulses. The ordinate values may pass through a simple A/D converter and then directly to the CPU in binary form without the need for character conversion and formatting. Which other spectrometer functions could be brought under computer control by the provision of the appropriate control lines through the interface (see also Sec. II.D.2.) is a matter for individual judgment.

B. Sampling Considerations

1. *Rate of Sampling*

The representation of a complex spectrum by a finite number of data points inevitably raises the question of what is an appropriate data density to use. It is important not to lose spectral information necessary for the particular analysis, but we should minimize the amount of data that has to be handled and stored.

As a rule of thumb, 10 or 20 points per half-height width of the narrowest band should suffice for most purposes. For organic liquids and solutions where half-widths are generally above about 10 cm^{-1}, an encoding interval of 0.5 cm^{-1} has been found satisfactory in both research and industrial environments [1,2]. Even at this density, a single full spectrum from 4000 to 200 cm^{-1} would require several thousand data points and may involve the punching and handling of several hundred feet of paper tape (Sec. II.C.3). Since the actual data density required is dependent on the nature of the subsequent processing, it is usual to have a range of sample intervals available, typically from 10 through 1 to 0.1 cm^{-1}. When checking the ordinate calibration of the spectrophotometer by the rotating sector method, for instance (Sec. III.A.1), it is clearly unnecessary to make measurements closer together than 10 cm^{-1} [1]. In precise studies of the shapes of narrow bands an interval of 0.1 cm^{-1} or even less may be advisable.

Figure 1 shows the progressive loss of spectral information as the sampling interval is raised from 0.4 through 1 to 5 cm^{-1}. At 0.4 cm^{-1} the loss is scarcely detectable. At 5 cm^{-1} the spectrum

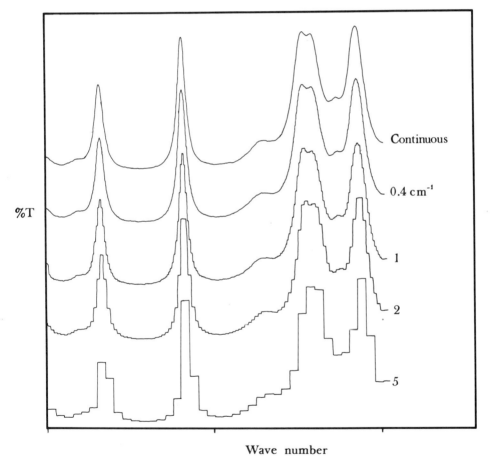

Wave number

FIG. 1. Part of an IR spectrum sampled at various abscissa intervals (0.4, 1, 2, and 5 cm^{-1}). (Reprinted by courtesy of the UV Spectrometry Group, U.K.)

looks so crude as to be useless. In fact, interpolative smoothing
can help considerably to restore a reasonable appearance (Sec.
III.B.1). In some quantitative analytical applications such a data
density may be entirely adequate (Sec. IV.C), and for simple comparative identification procedures it could be reduced even further
(Sec. VI.C).

The overall data acquisition time will be determined by the
scanning rate of the spectrometer (cm^{-1}/min) and possibly also by
hardware limitations of any mechanical digitizing equipment. The
latter problem is only likely to arise during off-line data logging
on punched paper tape. The punch rate of a TTY is typically 10
characters/sec; if ordinate-only data is recorded as "four digits
plus comma" blocks (i.e., five characters/record), the sampling rate
must not exceed 2 Hz. To correspond to a wave number interval of
0.5 cm^{-1}, the scan rate must therefore be held down to 60 cm^{-1}/min.
Much more flexibility is afforded by one of the high-speed punch
units, capable of 50 to 80 characters/sec [e.g., Digital Equipment
Corporation (DEC) model PC04; Facit model 4070] which allow fully
formatted, ordinate-only data to be recorded every 0.1 sec (Sec.
II.C.3).

In on-line systems, data acquisition can be very fast indeed
(several KHz is necessary; for interferometry, see Sec. II.B.4), although information theory indicates that nothing is gained by ex-
ceeding the spectrometer chopper rate. The only likely limitation
in practice is the amount of data that can be held in the memory,
unless provision is made for it to be automatically transferred to
disk or tape.

2. Precision of the Data

The precision with which intensity data are recorded need not
be much better than its uncertainty (i.e., the basic reproducibility
of the spectrophotometer itself or the noise level under the particular experimental conditions). This is typically about ±0.4% T with
the older null balance instruments, which corresponds to an ordinate
resolution of eight bits in the A/D converter [3]. Because it is
advisable to record one or two bits of noise, it is more common to

have nine- or ten-bit resolution available. Thus a 1,000 position
shaft encoder or a "3½ digit" DVM (0-1999) can record transmittance
to 0.1%.

For critical intensity measurements, a ratio-recording photometric system is clearly preferable, not only for the "live zero"
but also in terms of an improvement in reproducibility [4]. Even
so it seems unlikely that the precision of intensity measurements in
the IR is likely to greatly exceed the 0.1% T level that can now be
reached.

The precision of wave number encoding, if selected, depends on
the shaft encoder or stepper motor that is used in the prism or
grating drive, but is commonly to the nearest 0.1 cm^{-1} (1 part in
40,000). Wave number repeatability is generally better than ±0.5
cm^{-1}. Although the absolute accuracy will probably be lower, it may
be improved by calibration as necessary (Sec. III.A.2).

In data logging systems which sample ordinate only on a time
interval basis (Sec. II.C.3), the corresponding wave number values
are easily determined with an accuracy dependent on the synchronism
between the timer and the scan drive as well as the accuracy with
which the position of the first point is known.

3. Total Information Content

The total information content of a digitized spectrum is fixed
at the time and by the means of its acquisition. No amount of subsequent computation can increase the amount of true information contained in it. What we can achieve, often to considerable advantage,
is a redistribution of uncertainty between ordinate and abscissa by
means of noise smoothing and resolution enhancement (Sec. III.B). A
quantitative measure of the information content could be given by
the product of the ordinate and abscissa resolution in terms of a
total number of bits. For example, an ordinate resolution of 1 in
2^8 (0.4% T) and a position resolution of 1 in 2^{13} (0.5 in 4000 cm^{-1})
would give a total of 21 bits (approximately 2×10^6).

It often happens in difficult sampling situations (e.g., microsampling, strong background absorption, etc.), where the signal-to-
noise ratio (S/N) is unavoidably low, that the particular features

under examination disappear in the wake of oversmoothing. The only alternative is to increase the information content in the spectrum by remeasuring the data a number of times and combining the results (i.e., multiscan averaging, or CATing). This is unfortunately another process following the law of diminishing returns, since the improvement in S/N proceeds only as the square root of the number of scans. Nevertheless there are many difficult applications areas in analytical spectroscopy where a S/N improvement of even 10, achieved after 100 scans, is well worthwhile. A nice example is shown in the paper by Swalen [5].

Multiscan averaging is essential in FT spectroscopy, but the interferogram can be so rapidly remeasured that the extra data acquisition time is hardly significant (Sec. II.B.4).

4. *Operation in the Fourier Mode*

In optical FT spectroscopy [6], the experimental data essentially consist of the detected signal strength as a function of optical path difference. For a monochromatic source the interferogram produced is a cosine function whose period is determined by the source frequency. If several frequencies are present, the combined interferogram is the sum of the several cosine waves. As the complexity of the spectrum increases it rapidly becomes virtually impossible to mentally deduce anything about the form of the spectrum by inspection of the interferogram. This conversion is normally performed numerically, by computer program, using one of the versions of the fast Fourier transform (FFT) algorithm. The computer is therefore essential for the initial generation of the spectrum but, once incorporated into an instrument system, is also available for other applications programming. It does add considerably to the overall system cost, which has to be justified by the enhanced performance possible by operating in the Fourier mode.

In dispersive spectroscopy the detector is allowed to see only one resolution element at a time, as the spectrum is swept slowly across the exit slit position. Consequently, and especially at high resolution, measurement times may have to be long in order to achieve

an acceptable S/N ratio. In the Fourier mode, on the other hand,
all frequencies contribute simultaneously to the interferogram,
which can be recorded in a much shorter time (resulting in substantial time saving), and which is proportional to the square root of
the number of resolution elements. This is generally known as the
Fellgett (or multiplex) advantage. Also, because narrow slits are
not used to control the spectral resolution, the radiation beam
through the interferometer can be circular in section and a much
higher energy throughput is therefore available (Jacquinot's advantage). This too can be used either to increase the S/N or to reduce
the overall measurement time, but care should be taken not to squander
this advantage by noncircular sampling accessories.

Despite these performance advantages, the constraints of information theory still apply. In particular, the higher the resolution that is required in the final spectrum, the more extended must
be the interferogram. For the Michelson-type interferometer, this
means that the actual mirror travel must be increased accordingly.
Care must also be taken to sample the interferogram at a sufficient
data density, since this determines the highest spectral frequency
that can be properly recorded. In this context, it is important to
filter out source frequencies beyond the intended range lest they
masquerade as lower frequency components through undersampling (i.e.,
aliasing).

It is a characteristic of truncated time functions (e.g.,
interferograms) that their associated spectra show disturbing oscillations. One way of eliminating these is to multiply the interferogram by a function that smoothly reduces the amplitude to zero
at its upper limit. This process of apodization, as it is called,
is equivalent to a convolution (or smoothing; see Sec. III.B) of
the frequency spectrum and the imposition of a minimum line shape
function.

As far as the appearance of the final spectrum is concerned,
it is especially important that all phase errors are eliminated from
the interferogram, since their effect is to introduce a degree of

antisymmetry (i.e., high- and low-frequency wings raised and depressed in mutual opposition) that can be most misleading or even disastrous in complex spectra. Similar effects are also observed through missed or extra data points, since these inevitably upset the phase relationships of the data blocks.

These and other general aspects of optical FT spectroscopy have been discussed by various authors [6-10]. Although many practical implementations of Fourier operation have been devised, it is the rapid scanning Michelson-type interferometer that has so far made the greatest impact in the general IR field. An interesting alternative has recently been described by Decker [11], employing the Hadamard transform technique which features the multiplex advantage. It is claimed that a number of design attractions should result in greater simplicity and lower cost.

C. Off-line Data Logging

1. Indirect Data Entry

It was mentioned earlier (Sec. II.A.2) that a minimum entry system for computer-aided spectroscopy might consist simply of a data logger plus access to some computer facility, such as a time-sharing terminal. It should be added that in the 1960s, apart from a full on-line system to a medium-size computer (Sec. II.D.5), there was little alternative to logging the data on punched paper tape, having these tapes converted to magnetic tape for routine use by the computer center, with a final punching of a deck of cards after editing and error correction. This was the environment in which the pioneering work of Jones [1,12], Crawford [13-15], and many others was carried out. It suffered the great disadvantage that one or two big "runs" a day were all the throughput that could be expected. This meant, in particular, that program development time was painfully slow and the mechanics of handling and converting large volumes of experimental data were formidable.

Anyone seeking to minimize his capital outlay should by all means consider the use of a data logger, but also be prepared for the expense of rented computer time and data storage. It is surprising how easily this may exceed the initial hardware cost in even 1 or 2 years' operation. However, in some specific areas of quantitative or qualitative chemical analysis, where the volume of data transmitted is not great or where the computation is not especially complex or time-consuming, this route can be most economical. It is, nonetheless, somewhat restrictive. A full spectrum of a few thousand points seems to take an age to transmit via a TTY. With some exceptions (mainly on private in-house installations), time-sharing systems are not noted for convenient graphic output facilities. However, with the continuing spread of magnetic tape cassette units and visual display terminals, the situation may well improve, but so too can the capital outlay rise.

It should be clear from the preceding paragraphs that the problem has much to do with computers that are remote from the laboratory and little to do with off-line data logging per se. Equal consideration should therefore be given to the stand-alone minicomputer, with its own high-speed tape reader, as a data processing system in the laboratory, in many ways as useful as and more convenient than a larger remote center. There must be a large number of such machines (e.g., DEC PDP-8 and PDP-11, Data General Nova, and many others) already installed for the processing of data logged off-line from several different instruments.

For the rest of this section, we shall assume that some appropriate computing facility is available, and consider the relative merits of the various off-line data storage media, as well as some examples of data-logging systems that use them.

2. Punched Cards

Although not strictly speaking a data-logging medium, punched cards are universally used for data input to large systems and are a long-term data storage medium. Nevertheless, Biggers [16] has described the use of data acquisition from two Cary 14 spectrophotometers with punched card output, although no details are recorded.

Perhaps the most obvious convenience features are the instant readability of the card contents (once "interpreted") and the ease of editing by simple exchange. The danger lies in the risk of

accidental misplacement. It is always advisable, therefore, to include in any programs using card input a sequence-checking subroutine.
This is a feature of the three volumes of programs issued by Jones
and co-workers at the National Research Council of Canada (NRCC)
[17-19], which we shall discuss in more detail in later sections.
Sequence checking requires that the wave number value (or wavelength)
be recorded with each ordinate. This reduces to eight the number of
records that can be held on a single 80-column card [each containing
five digits of wave number to 0.1 cm^{-1} and four digits of ordinate
to 0.1% T, as in the Fortran format 8 (I5,I4,1X)]. It would therefore take nearly 700 cards (about a third of a "standard box") to
contain the 5,500 records of a full-range spectrum (4000 to 2000 cm^{-1}
every 1 cm^{-1}; 2000 to 250 cm^{-1} every 0.5 cm^{-1}). The same data as
ASCII characters (four digits of ordinate only, plus comma as separator) would require about 200 ft of punched paper tape (about a 2 in.-
diameter reel), or barely 10% of the capacity of a single magnetic
tape cassette.

3. Punched Paper Tape

Punched paper tape is the traditional off-line medium and is
characterized by medium cost, medium speed, visual checking, and
simple manual editing. It can be used with virtually every computer
installation, although the larger computer centers prefer to convert
to magnetic tape or cards for subsequent input of the same or corrected data. High-speed paper tape readers for minicomputer systems
typically operate at 300 characters/sec. The full range spectrum of
27,500 ASCII characters (previous section) could therefore be read
in about 90 sec (cf. 45 min via TTY). If the wear and tear of repeated use is troublesome, a copy of valuable data tapes should be
made on Mylar, which is virtually indestructible.

In the early system developed by Jones [1], a Perkin-Elmer
Model 521 spectrophotometer and Digital Data Recorder unit are used
in conjunction with a Teletype Model BRP-E11 eight-track 1 in.-paper
tape punch. Each nine-character record represents five decimal
digits of abscissa (wave number x 10), three digits of ordinate

(percentage transmission x 10), and a single end-of-record character. Since a four-bit binary code plus one track for parity (odd) was used instead of ASCII, tapes are listed when required on a Flexowriter (Friden Corporation). Numeric control data such as sample identification number, end-of-run word (e.g., 99999999), could be entered manually character by character. This was also useful to separate the main spectrum from ordinate and abscissa calibration data (Sec. III.A.), the whole being kept as a single length of tape.

Sampling intervals of 10, 5, 2, 1, 0.5, 0.2, and 0.1 cm^{-1} are available, although the maximum punching rate will impose a limitation on the spectrophotometer scan speed. Provision is also made for an expanded chart record on a slave recorder, with fiduciary marks along each edge triggered by the encoding system for direct comparison with the punched tape.

The NRCC system of Jones [1] described here was the forerunner of the ADS I and II units (Perkin-Elmer Corporation), which differ only in whether or not the wave number value itself is included in the tape record (ASCII). In the ordinate-only version (ADS II), the sampling interval is still selected in terms of wave number but a trigger commutator is used to activate a single transmittance encoder. In this way accurate wave number sampling is guaranteed, although the actual abscissa values must be determined by counting from the first point.

A similar system (i.e., ordinate-only recording) is used in the Data Punch unit (Perkin-Elmer Ltd.) which samples on the basis of a selectable time interval. The time base is taken from the mains frequency and the unit is therefore suitable for use with any mains synchronous scanning instrument. A 3½ digit DVM samples the ordinate as an analog voltage, which is then converted into an ASCII character string (with a range of formatting options), with final output to a high-speed paper tape punch (Facit Model 4070). On the Perkin-Elmer X57- and X77-type IR spectrophotometers, sampling is initiated by the wave number marker, which ensures the positional accuracy of the sampled points and is inhibited during the filter

and grating changeover periods.   The sampling rate may be varied
between 0.1 and 10 Hz.   The unit is illustrated in Fig. 2.

Larsen [20] has described in detail the construction of a low-
cost digital data logger which can record one reading per second,
using the tape punch of a standard TTY.   Analog voltage data are

FIG. 2.   Paper tape punch (Facit Model 4070) and control unit for
converting analog voltage signals into four-digit ASCII character
records, sampled at selectable time intervals.   (Reprinted by courtesy of Perkin-Elmer Ltd., U.K.)

input to a DVM unit (Keithley Model 160) which performs the initial
A/C (Analog/Digital) conversion to BCD code (for the display digit
drivers), and a BCD to ASCII code coupler and formatter is connected
to the TTY for printout and simultaneous punching. Applications to
IR spectroscopy were implied but not described in detail.

4. *Magnetic Tape System*

The electronic read/write feature of magnetic tape systems
greatly reduces the data rate limitation imposed by the mechanical
components of paper tape punches. A much higher data-recording
density is possible and, in distinct contrast to paper punches, they
are silent in operation. The so-called industry-standard 1/2-in.
tape systems are available in seven- or nine-track versions, with
the data in bit parallel format across the tracks, at a longitudinal
packing density of between 200 and 800 bits per inch (bpi). Therefore a 1200-ft reel of such tape can accommodate over 10 million
characters or up to 1 million records of spectroscopic data. A
more economical and very convenient alternative is the DEC tape
system (Digital Equipment Corporation), using 10 cm diameter reels
of 3/4 in. tape with data density of about 350 bpi and a total capacity of over 300K characters. More recent still are the "Phillips"
and "3-M" type of magnetic tape cassette units, which offer outstanding convenience and economy but still with very good data storage capacity and transfer rates.

Of the commercially available off-line data systems for IR
spectroscopy, the Perkin-Elmer Model 180/DMR (digital magnetic recorder) configuration is worth a brief description here, if only to
illustrate the level of performance that is possible in off-line
data logging. A typical installation is shown in Fig. 3. Of particular interest is the digital nature of the primary ordinate and
abscissa data resulting from the digital ratiometer system and stepper motor wave number drive, and the precision which is retained
through the interface. The standard output (ASCII characters) is
presented as six digits of abscissa (wave number to 0.01 cm^{-1}) and
five digits of ordinate (to 0.01% T) with appropriate unit and

FIG. 3.   IR spectrophotometer with digital magnetic recorder system
for off-line data logging. Values of ordinate (to 0.01% T) and
abscissa (to 0.01 cm^{-1}) are sampled at intervals ranging from 10 to
0.01 cm^{-1}, and stored as ASCII characters on seven- or nine-track
tape. (Reprinted by courtesy of Perkin-Elmer Corp., U.S.)

record separator characters. A single control allows selection of
the data-sampling interval from 10 to 0.01 cm^{-1}.

The magnetic tape system uses 1,200-ft reels of seven- or nine-
track industry standard tape. A flexible control system is provided
which, in conjunction with a TTY, allows the following modes of
operation: initialization data entered at the TTY keyboard to be
transmitted to tape; data logged from the spectrophotometer to be
recorded directly on tape, or printed at the TTY for inspection, or
both (provided the data rate is not too high for the printer); automatic shutdown if the data rate (combination of scan speed and wave
number interval) exceeds the limits of the tape system. Also featured are optional and variable length inter-record gap generation
with buffering of data logged during this period. Maximum character

rates are 150 or 500 Hz, respectively, with or without inter-record
gap generation. The final product is therefore a fully formatted
1/2-in. tape that may be directly mounted on any large computer
system.

At the other end of the scale are the compact, economical,
general purpose data-logging systems that use magnetic tape cassettes.
Typical of several recently announced is the Model AL-4 (Digitronics
Ltd., Milton Keynes, England). It is basically equivalent to the
Data Punch unit (Fig. 2) in that several voltage ranges and sampling
rates are available (0.1 to 16 sec interval), although it must be
remembered that an equivalent cassette reader (or replay unit) will
be required before the recorded data can be entered to a computer
system.

Mattson and McBride [21] have described an interface for a
Hilger-Watts H-1200 grating spectrometer and a Mobark 305 cassette
tape recorder. Ordinate data are recorded in eight-bit binary format
(transmittance to 0.4%) every 2.5 cm^{-1} (4000 to 2000 cm^{-1}) or 0.5 cm^{-1}
(2000 to 650 cm^{-1}) using a photodiode and Schmidt trigger to sample
the data once every 10 revolutions of the chart drive motor. This
corresponds to a data rate of three points/sec, which could be increased by a factor of 10 by sampling every motor revolution and
still be within the incremental speed limit of the cassette recorder
(120 characters/sec). A single 300-ft cassette tape can hold up to
42 spectra of 3,300 eight-bit records at the above data density.

D. On-line Spectrometer Systems

1. Direct Data Entry

The relative merits of on-line and off-line operation will no
doubt continue to be debated for some time to come but we should be
clear about the issues. It seems appropriate here to ask the following question: if a laboratory has an IR or Raman spectrometer
and a modern minicomputer system, is it worth interfacing them together? If permanent digital records of spectra are to be retained
in a medium and using the equipment that could have been used for

off-line data logging, then the expense of a special on-line interface could be saved. It should be recognized that with the present
variety of spectrometers and computers the manufacture of special
interfaces must suffer from the high cost of all low-volume products.
In the general case, then, the discussion is reduced to one of cost
and convenience.

But, as we shall wee in some of the examples that follow, the
design and construction of a simple data-logging interface is by no
means beyond the capability of many present day laboratories.
Spector [3] has discussed some of the factors to be considered, A/D
conversion techniques in particular. Those unfamiliar with the
characteristics of integrated circuit devices would do well to con-
sult the report by Dessy and Titus on computer interfacing [22].
General purpose data acquisition systems have been described in
some detail by Lauer and Osteryoung [23] and Ramaley and Wilson [24].

A simple means of data logging direct to a remote computer
center has been described by Krag and co-workers [25] using a DVM
and code coupler (i.e., similar to Larsen's paper tape logger [20],
Sec. II.C.3). The data are directly transmitted over a standard
phone line to an IBM 360/40 system where they are automatically recorded on magnetic tape for subsequent processing on demand from a
terminal.

The indications are, then, that a simple interface for data
logging only can indeed be cheaper than the paper tape punch or magnetic tape recorder it replaces. Furthermore, as the price of minicomputers themselves continues to fall and their numbers to grow, it
seems inevitable that on-line spectroscopy will become ever more
commonplace.

2. *Real Time Control*

It is where some degree of useful control of the spectrometer
can be exercised by the computer that we begin to see the real justification of on-line operation as opposed to its mere convenience
value. Nevertheless, it is hard to see in what respects the performance of most IR spectrometers produced today could be much

improved by computer control. The analyst should still make the choice as to what resolution is required, what noise level is acceptable, and how long he is prepared to wait for his result.

However, two situations come to mind where computer control would be valuable; namely, for unattended or remote spectrometer operation (perhaps because of hazardous samples or environment), and with automatic multisampling systems where the various samples require different spectrometer settings and data analysis procedures. If on-line data logging were considered desirable, it then again becomes a matter of operator convenience for the computer to have access to the front panel controls of the spectrometer, if only so that all human interaction may be via a single means.

The question of real-time control (dynamic instrument readjustment for optimization of data acquisition, or in response to a recognized malfunction) seems unlikely to arise in conventional IR spectroscopy. In interferometry, on the other hand, quite the opposite is true. In particular, accurate, active control of the moving mirror drive is critical for successful operation of the instrument (Sec. II.D.4). As we saw in Sec. II.B.4, there are many special considerations in FT spectroscopy that make on-line operation and control essential features of the technique.

3. Calculator-Controlled Systems

Sakalys [26] has described a spectrophotometer data processing system controlled by a programmable calculator (Model 700A, Wang Laboratories). Although the example concerns the visible region, it could also provide a useful example for a simple IR system. The principal attraction is economy; the simplicity of the interface hardware allows for easy modification and extension, but at the same time one is somewhat restricted in the complexity of software that can be generated. Nevertheless, the successful peak resolution (by fourth differences) shown by Sakalys is a good demonstration of the possibilities of such an approach. Klauminzer [27] has developed a similar spectrometer system using a Hewlett-Packard-type 9100-A desk top calculator for data acquisition and simple control.

Electronic calculators make considerable use of large-scale
integration (LSI) components to perform particular arithmetic functions. In this respect they are similar to the very low-cost microprocessors that are now available and which are suitable for relatively
simple control and data-processing applications. The only distinction
that now seems possible between calculators and computers (e.g.,
microprocessors) is that data transfer and logic operations are
typically serial in the former case and parallel in the latter.
Reilley [28] has described a very compact unit that acts as a front-
end preprocessor to a time-sharing system for on-line data acquisition
from a number of laboratory instruments. The central processor is a
single integrated circuit (Model 8008, Intel Corporation) which can
be linked to memory modules, A/D converters, character display and
keyboard, or tape cassette drive. In this way, a very cheap micro-
computer system may be built up to perform any of a wide range of such
laboratory functions. The principal obstacle to be overcome in a
microprocessor system is the difficulty of developing and testing the
software program.

4. *Computer-based Spectrometer Systems*

We shall consider here some examples described in the literature
or which are commercially available, where IR or Raman spectrometers
have been interfaced to medium-size or laboratory-based computers, and
leave to the following section a discussion of larger multi-instrument
systems. The distinction between the two categories is often only one
of emphasis, or detail in reporting, and is made here mainly for convenience.

At the 1968 Eastern Analytical Symposium (Atlanta, Georgia),
Chuang and co-workers [29] described the setting up and initial operation of a Perkin-Elmer 621 spectrophotometer coupled to an IBM 1800
computer system. Shaft encoders were used to provide ordinate and
abscissa information (three and five digits, respectively) which was
transmitted as 32 bits of BCD code (8 x 4 bits per digit) at a rate
of up to about 10 data points/sec. After receiving a "ready" message
at the terminal in response to an operator request to initiate data

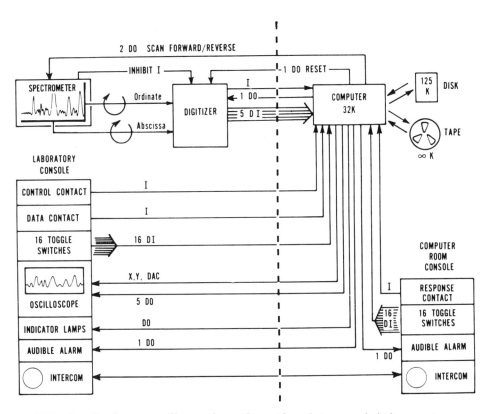

FIG. 4. Hardware configuration of on-line data acquisition system
for IR and Raman spectroscopy. The dashed line separates equipment
based in the laboratory (left) and computer center (right). DI is
digital input, DO is digital output. (Reprinted from Scherer and
Kint [30], p. 1616, by courtesy of the editor of *Applied Optics*.

acquisition, the spectrometer scan was started manually and data
transmitted until finally the interface was switched off-line. The
eight-digit BCD input was converted to two 16-bit words of ordinate
and abscissa, and transferred to disk a sector at a time (320 words)
in a Fortran-readable format.

An IBM 1802 computer (with 32K of 16-bit core plus disk and tape
backing store) was used by Scherer and Kint [30] for the development
of a combined IR and Raman system. The total hardware configuration
is shown diagrammatically in Fig. 4. A Cary-White 90, ratio-recording

IR spectrometer and two Raman instruments (Spex 1401 photon counting
and Perkin-Elmer LR-1 synchronous detection) were interfaced to a
Cary-Datex digitizer system. This has an ordinate encoder range of
-50 to +1050 (giving a resolution to 0.1% with limited overshoot)
and abscissa encoding to 0.1 cm^{-1} (or "drum number"). Data are transferred from an instrument interface to core either on an interrupt
basis or by an automatic cycle stealing process, with data rates of
up to 50 points/sec. Real-time parity and sequence checking is done
in core, using two buffer tables alternately, before the data is
transferred to disk as 320-word sectors. A high degree of interactive
control and operator communication is provided (using a visual display
screen and an intercom line), which contribute significantly to the
overall success and versatility of the system.

Shapiro and Schultz [31] have discussed in detail the design of
a laboratory-based computer system (using a Honeywell DDP-516 with
32K core, 1.8M word disk, and a wide range of peripherals) which
services several different on-line spectrometers. Each is connected
to a remote operator's console, which houses the basic interface,
plus status/control and parameter entry panels, display oscilloscope,
and intercom to the computer room. The interface to a Beckmann IR-7
spectrophotometer uses a 100-turn, 100-count shaft encoder (to 0.1
cm^{-1}) outputting in Gray code for the abscissa, and a linear strip
encoder for recording absorbance. A Cary 81 Raman spectrometer was
fitted with a 100-turn, 1000-count shaft encoder, corresponding to
a resolution of 0.04 cm^{-1} over the 0 to 4000 cm^{-1} range.

Probably the best known commercially available computer-based
IR system is the Model FTS-14 produced by Digilab Inc., Cambridge,
Mass. This is a complete hardware/software packaged system, which
combines a Model 296 interferometer with a Nova minicomputer (Data
General Corporation, Southboro, Mass.). The high-resolution, rapid-
scan Michelson-type interferometer can be operated in either single-
or double-beam mode, using a coated potassium bromide (KBr) beam
splitter and fast response triglycine sulfate (TGS) pyroelectric
bolometer as detector over the 3800- to 400- cm^{-1} frequency range.

With an appropriate accessory, involving a change of beam splitter, detector, and source unit, the far IR region is accessible down to 10 cm^{-1}. Presample modulation and digital ratio recording contribute significantly towards the high photometric accuracy that is attainable.

The minicomputer system is available in two configurations: all-core (4 to 16K 16-bit words) or core plus disk (128K, expandable). The schematic diagram of the disk system is shown in Fig. 5. The computer not only handles the high-speed acquisition of the inter-ferogram data but also, in conjunction with a laser-driven reference interferometer, accurately controls the mirror movement during each scan. Complete control is exercised over the basic operating param-eters (resolution, number of scans, etc.), which are entered by the operator at the TTY keyboard. The principal data processing is the generation of the spectrum from the interferogram, using a modified FFT algorithm, with appropriate apodization and phase correction. Having the digital spectrum already in core, it is immediately avail-able for any subsequent processing that may be required. A range of support software is provided for replotting, addition, subtraction, and other functions [32].

5. Multiple Instrument Systems

In considering multiple instrument/computer systems, we are in danger of trespassing outside the scope of the present chapter into the much wider field of laboratory automation in general. An excel-lent review of the general applications of computers in the chemistry laboratory has been given by Perone [33]. The article by Swalen [5] gives a good idea of the scope of computer assistance in various branches of spectroscopy. Complex systems based primarily on high-resolution MS, nuclear magnetic resonance (nmr), and GC have been described by Poppe [34], Hallett and co-workers [35], and Elkins [36].

Most systems of this kind have been built up around one or two key instruments for which on-line computing is of primary importance in the handling of very high data rates (MS, GC-MS) and in FT data processing (IR interferometry and pulsed nmr). In this situation, the extension of the system to dispersive IR or Raman equipment is

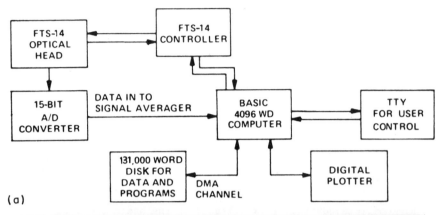

FIG. 5. On-line interferometric IR data system, Digilab Model FTS-14 with 131K word disk store: (a) schematic diagram; (b) typical installation. (Reprinted by courtesy of Cambridge Scientific Instruments Ltd., U.K.)

all too often relegated to the level of desirable [36], but hardly essential. Nevertheless, the LABCOM system described by Deane and co-workers [37] gives IR and Raman a prominent place and uses CAMAC standard interface hardware between the laboratory instruments and

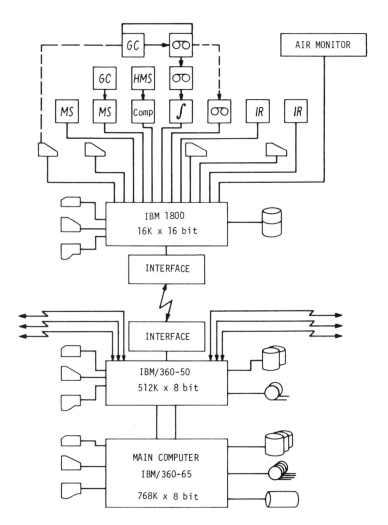

FIG. 6.   Schematic diagram of the analytical laboratory computer
system at Badische Anilin - & Soda-Fabrik AG, Ludwigshafen, Germany.
(Reprinted from Günzler [2], p. 880, by courtesy of the editor of
Chemie-Ingenieur-Technik.)   See text (p. 378).

a PDP-8/I (Digital Equipment Corporation) computer. The University
of Oregon Chemistry Department system outlined by Klopfenstein [38]
has provision for both IR and Raman facilities, and is especially
interesting as an example of the newly emerging hierarchical computer
systems.

Finally, it seems fitting to close the present section with a
brief description of the impressive system installed by B.A.S.F.
(Badische Anilin- & Soda-Fabrik AG, Ludwigshafen, Germany), as described by Günzler [2]. A schematic representation of the overall
hardware configuration is shown in Fig. 6. The system includes five
IR spectrophotometers [Perkin-Elmer Models 125 (2), 225, 021, and
157], of which any two may be on-line simultaneously, interfaced to
an IBM 1800 (16K x 16-bit words of core), which is itself acting as
a front-end communications processor to a remote IBM 360/50 system
in time-sharing mode. This is further connected to an even larger
IBM 360/65 system in the background. It is interesting to note that,
from an IR spectroscopic standpoint, one of the major features of the
computer system is that it provides a fully automated means of quantitative chemical analysis, including Beer-Lambert plots and a formatted
printout of peak positions, heights, and areas. The chemist, who
requests a particular analysis with the sample ready in the spectrometer, receives in return from the computer a full analytical report
ready for filing.

III. DATA REFINEMENT

A. Calibration and Formatting

1. *Ordinate Calibration by Rotating Sectors*
The question of photometric accuracy in IR spectroscopy has
been investigated by Stewart [39] and by Jones [1,4,40], who has
developed a simple but accurate method for the routine calibration
of digitally recorded spectra. The problem of nonlinearity is thought
to arise mainly from mechanical imperfections in the comb attenuators
characteristic of null balance double-beam instruments, but is also
affected by the focusing and alignment of the optical system. A set
of eight sector photometers (originally obtained from Research and
Industrial Instruments Co., London, England, but now marketed by
Beckman Instruments, Inc.) is used to cover the transmittance range
from about 6 to 48% T. These small, high-speed rotating sectors

were originally calibrated against another accurately standardized
reference set. Now a simple, white-light photometer has been set up
for the periodic checking of sector transmission values to an accuracy
of ±0.05% T [4].

The nature of the error curve (true vs. observed transmittance)
is determined simply enough by running some or all of the sectors in
the wave number range of interest. If this is basically linear, then
simple interpolation between two convenient values (say, 20 and 80% T)
will suffice to calibrate the whole scale. However, for the Perkin-
Elmer 521 spectrophotometer (null-balance-type), Jones found that an
approximately parabolic correction curve was obtained [1]. For the
most accurate intensity measurements, it would therefore seem advisable to calibrate the ordinate scale at intervals of approximately
10% T, at least around the intensity value of the actual sample spectrum, and to carry this through the whole wave number range used
(e.g., every 10 cm^{-1}). The calibration program should then include
provision not only for interpolating the true sample intensity from
the true and observed sector data, but also for interpolating the
appropriate sector data to use for wave number positions within these
10-cm^{-1} intervals.

From a statistical analysis of sector measurements made over the
course of 1 year [4], it is clear that an absolute ordinate accuracy
of about ±0.4% T is obtainable. Preliminary results over a shorter
period made on a ratio-recording instrument (Perkin-Elmer Model 180/
DMR) and using the same sector calibration procedure, show an improvement by a factor of 2 on this figure [4].

2. Wave number Calibration

Wave number tables for the calibration of IR spectrometers have
been published [41] and contain accurately measured, internationally
agreed values of band positions for a range of vapors and liquids,
covering the whole IR region (4000 to 280 cm^{-1}). The spectrum of
liquid indene has been recommended [1] as a satisfactory general
purpose wave number standard, using several key bands to define the
spectrometer error curve. In studies over a limited spectral range

where high accuracy is required, it may sometimes be more appropriate to use instead the vibration-rotation lines of a suitable
vapor.

A convenient procedure is as follows. After the prime sample
spectrum has been recorded, the calibration standard spectrum is
digitized at a suitable data density. The computer program must
first locate the absorption maxima of the measured calibration spectrum using some peak-finding routine (Sec. IV.A.1) and construct a
data table of wave number error vs. nominal wave number value using
stored standard values for the bands measured. The sample spectrum,
a list of wave number and ordinate values, is then corrected by
subtracting the appropriate error value from every point. It is
usually adequate to use a linear interpolation for intermediate
positions.

The calibrated spectrum thus obtained will now consist of the
set of ordinate values (perhaps themselves already corrected by the
sector data) at their new abscissa positions, and will no longer be
equally spaced by the original sampling interval. The final part
of the calibration procedure is therefore to generate a new set of
ordinate values at the original wave number positions, again using
some interpolation subroutine. Since the wave number correction is
likely to be not insignificant compared with the sampling interval,
it is preferable to use quadratic (or cubic) interpolation (Sec.
III.B.1) to preserve the ordinate accuracy.

3. Scale Conversion

Two of the most obvious conveniences of having a spectrum in
digital form are the ease with which it may be redrawn in any shape
or size on appropriate plotting facilities (via a D/A converter to
a pen recorder or oscilloscope) and the ability to change the units
of the data, for example, from transmittance to absorbance, or wave
number to wavelength. These facilities can be particularly useful
for documentation or comparison purposes; Anderson and Woodall [42]
give a good example of the dramatic transformations that can result.
A general purpose Fortran program is available for these requirements
[12,18].

B. Smoothing and Sharpening

1. Interpolation

We have already mentioned (Sec. III.A) several uses of interpolation in the accurate scale calibration of digitally recorded
spectra. A more general requirement might also be found in the
manipulation or storage of spectra at different data point densities.
In the simplest case of linear interpolation, between the two points
whose coordinates are x_1, y_1 and x_2, y_2, to find the ordinate value
y at an intermediate point x, we immediately recognize the expression

$$y = \frac{y_1 + (y_2 - y_1)(x - x_1)}{(x_2 - x_1)} \tag{1}$$

As soon as we find the need for quadratic, or higher order,
interpolation, the complexity of the equations begins to escalate.
Nevertheless, the solution can be generalized in terms of a system
of simultaneous linear equations. It is well known that three
points will serve to define a quadratic, four a cubic, five a quartic, and so on. We must first determine the coefficients of the
polynomial of order m that is defined by the m + 1 points chosen.
Thus for a cubic, defined by

$$y = a_0 + a_1 x + a_2 x^2 + a_3 x^3 \tag{2}$$

we may write the system of simultaneous equations:

$$y_i = a_0 + a_1 x_i + a_2 x_i^2 + a_3 x_i^3 \tag{3}$$

where i = 1, 2, 3, 4. This may be solved by standard procedures to
yield the vector of coefficients a_0, a_1, a_2, a_3, which may then be
used to calculate the value of y for any x. It will be noted that
Eq. (1) follows immediately from Eq. (3) for the case where i = 2.
In this general approach to interpolation, there is no requirement
that the input points be equally spaced along the abscissa. If,
however, they are so spaced, various procedures are available for
direct calculation of the result without the need to solve Eq. (3).

For example, a five-point Legrange polynomial function is used in the scale conversion program mentioned above [12,18].

Interpolation can also be achieved by a convolution process (Sec. III.B.2) and can be used to smooth out the irregularities of coarsely sampled spectra (e.g., Fig. 1). A new input array (of ordinate values) is taken, repeating each point as required across the original sampling interval. The effect of convolution (digital filtering) in this case is to replace the "steppiness" of the original data with the smoother profile of the convolution function.

2. Noise Smoothing

Noise smoothing in IR spectroscopy may become necessary in various situations (microsampling, strong background absorption) where the S/N ratio cannot be raised without unduly prolonging the measurement time. It should be regarded as a means of seeing better one feature (e.g., peak height) at the expense of another (loss of multiplet resolution or band broadening). No new information about the spectrum is added; the uncertainties are merely redistributed.

A real opportunity for improvement does exist, however, if the filter circuit of the spectrometer, with its nonideal (and always nonsymmetric) characteristics, can be bypassed during data acquisition and replaced by subsequent digital filtering in the computer. In 1964 Savitzky and Golay [43] proposed a set of polynomial functions for smoothing spectra in this way by numerical convolution. Their tables have been adopted as a standard ever since (with modifications suggested by Steinier and co-workers [44]), although generally without any formal guidelines for selecting the most appropriate set. Rules for their application have been determined empirically. Jones and co-workers [45,46], for example, have advocated that five- and seven- point quadratic functions be used, either singly or in combination, since the distortion introduced thereby is less than with the more extended sets. The nine-point quartic function has also been found successful and is featured in the NRCC programs [18].

It will be realized, however, that such a discussion is meaning-
less without reference to the wave number interval of the sampled
spectrum, the half-width of its absorption bands, and the degree of
noise smoothing required. We should also be prepared to specify the
amount of distortion that is acceptable (in terms of peak attenuation
and broadening). These aspects of spectrum smoothing have recently
been discussed by Porchet and Günthard [47], Yamashita and Minami
[48], and Cram and co-workers [49]. The very reasonable suggestion
has been made [48] that the peak attenuation should not be allowed
to exceed the output noise level, if no information is to be un-
necessarily discarded. The same authors have formulated conditions
that meet this criterion, in terms of the experimental parameters
(band half-width, sampling interval, etc.), and have given a graphical presentation [48] of the relative errors in peak height and
band half-width for noise-equivalent filters of polynomial degree
0 to 6.

These results clearly show that the higher the degree of polynomial used, the less the attenuation within the pass band and the
sharper the cut-off into the noise region. However, little is really
gained by going much beyond the quartic. In Fig. 7 the frequency
response curves are shown for the various filters whose impulse response represents the convolution function. The S/N improvement
that is achieved is proportional only to the square root of the
number of points taken. Beyond a certain stage there is no alternative but to increase the real information content of the data by
multiscan averaging (Sec. II.B.3).

The process of convolution is in fact no more than a weighted
running average and has been thoroughly treated by Bracewell [50].
A general discussion of the digital filtering of spectra has been
given by Ormsby [51]. In the Fourier-related time domain, an identical result is achieved by multiplication of the transforms of the
two functions (i.e., of the data and filter function). Likewise,
as we shall see later (Sec. III.B.4), deconvolution may be achieved

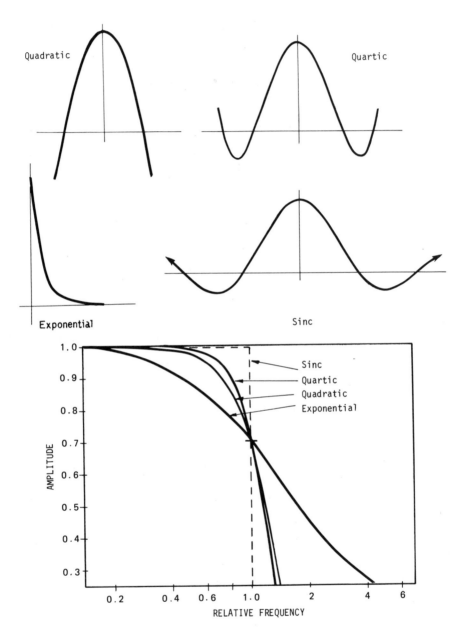

FIG. 7.   Impulse response (amplitude vs. time, above) and frequency
response (amplitude vs. frequency, below) characteristics for exponential (e.g., single RC circuit), quadratic, quartic, and sinc
function digital filters.

by division of their FTs. Finally, it is worth noting that the
repeated use of a single filter (or various ones in sequence [45,
46]) is exactly equivalent to a single convolution of the data, with
the function produced by first convoluting together the filters
themselves.

3. Resolution Enhancement by Pseudodeconvolution

One obvious way of improving the S/N level is to use a wider
slit program at the time of data acquisition. Unfortunately, this
will inevitably lead to a broadening of narrow bands and a loss of
multiplet detail. It is natural to inquire whether any degree of
resolution enhancement can be obtained by the use of a suitable
computer program.

An iterative procedure has been developed by Jones and co-
workers [45,46] and gives good results when the original distortion
is not too large. The method is termed pseudodeconvolution, because
the result should only be regarded as a good approximation to the
true deconvoluted spectrum. The argument is as follows. The data
recorded by the instrument are really a convolution of the true
sample spectrum with the slit transfer function of the monochromator,
assuming, of course, that it is recorded sufficiently slowly that no
distortion is added by the smoothing circuit of the signal electron-
ics. If we know, accurately, the shape and width of the slit function, we may convolute the recorded data a second time, numerically,
in the computer. Since we would expect the original distortion
(i.e., attenuation and broadening) to now be roughly doubled, we
can reconstruct an approximation to the real spectrum by subtracting
this added distortion from the recorded spectrum. We can check
whether this is, in fact, correct by further convoluting the result
and comparing it with the data actually recorded. In practice,
several cycles of iteration are required before this check is satis-
fied. It should be noted that if the process is continued indefinitely, oscillations are introduced which eventually dominate the
result [45]. This is only to be expected, however, in the light of
the fact that the total real information content of a spectrum is

limited, on the one hand by the abscissa resolution, which we are
here seeking to increase, and on the other hand by the noise level
of the ordinate, which we must be prepared to amplify.

Even where the experimental S/N is good, oscillations can be
introduced by round-off errors in the computation before the true
spectrum has been recovered fully. Jones and co-workers [45] have
used five- and seven-point quadratic smoothing functions to prevent
these effects becoming troublesome, but have warned of the dangers
of the indiscriminate use of such measures. They have also given
[46] a detailed analysis of the residual errors in the final result.
A Fortran program is available as part of the NRCC collection [18].

We should not lose sight of the fact that such a technique can
only be used if the slit function itself is known with reasonable
accuracy. Although it is widely assumed to be triangular with a
half height-width equal to the nominal "spectral slit width," this
is unlikely to be an exact representation [52]. Jansson and co-
workers [53] have described a method of measuring it at very high
resolution using one of the rotational lines of the N_2O band at
2798 cm^{-1} as a (virtually) monochromatic source. Jansson has described in detail [54] the generalized method of deconvolution that
was used and for which the procedure followed by Jones and co-workers
[45] appears as a special case. Figure 8 shows the degree of enhancement that can be achieved by this method [55]. A more widely
spaced set of lines occurs in the spectrum of HBr [41], which may
be useful in the conventional resolution range of about 1 cm^{-1} and
for which the isotopic splitting is negligible by comparison. A
laser source, while being highly monochromatic, is less suitable
for this purpose on account of its high degree of coherence.

A technique that has often been used for resolution enhancement,
particularly in nmr spectroscopy, derives from the empirical observation that if some fraction of the second derivative is added to
the recorded spectrum, the result appears considerably sharpened.
The theoretical justification of such an approach has been investigated by Steele and Hill [56], but it should be remembered that the
sharpening effect of the second derivative is accompanied by its
own higher noise level.

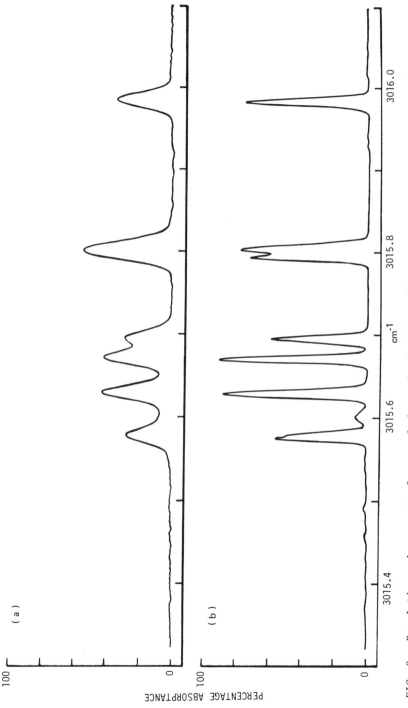

FIG. 8. Resolution enhancement of part of the ν_3 band of CH_4, (a) before and (b) after iterative deconvolution. (Reprinted from Jansson, Hunt, and Plyler [55], p. 598, by courtesy of the editor of *Journal of the Optical Society of America*.)

4. Deconvolution via the Fourier Domain

It has already been mentioned (Sec. III.B.2) that the convolution of spectra and filter functions is equivalent to the multiplication of their FTs. This is illustrated in Fig. 9 for the self-convolution of the rectangle function whose FT is the (sin x)/x or sinc function [50]. It will be remembered (Sec. III.B.2) that the ideal frequency response for a low-pass filter is rectangular, and its Fourier pair (the sinc function) is therefore the ideal smoothing convolute. It is equally true that a rectangular smoothing function (i.e., simple moving average) has a frequency response given by the sinc function, causing serious attenuation at quite low relative frequencies.

When the two sinc functions on the left of Fig. 9 are multiplied together, the product is, not surprisingly, the function $(\sin x)^2/x^2$, which is the FT of the triangle function. It is a simple matter to show that this result is indeed identical to the convolution of the two rectangle functions [50]. If we follow the diagram the other way around, that is, if the triangle is deconvoluted with one of the rectangles by dividing their transforms, the quotient is the transform of the other rectangle.

It is instructive to show another example of this process, since we shall later have cause to refer to some of the properties of the functions represented. Figure 10 illustrates the deconvolution of a two-stage RC filter (double exponential) into its two first-order components. Firstly, we notice that the transform of the exponential function is a Cauchy (Lorentz) function, a result we shall use later (Sec. V). Secondly, the transform has both real and imaginary components. It is a fundamental property [50] of any asymmetric real-only time function (e.g., the exponential impulse response) that its FT has a symmetric real component and an antisymmetric imaginary component (such functions are said to be Hermitean). Its importance in the present context is that we recognize the need for complex multiplication and division for the correct implementation of convolution and deconvolution in the Fourier domain.

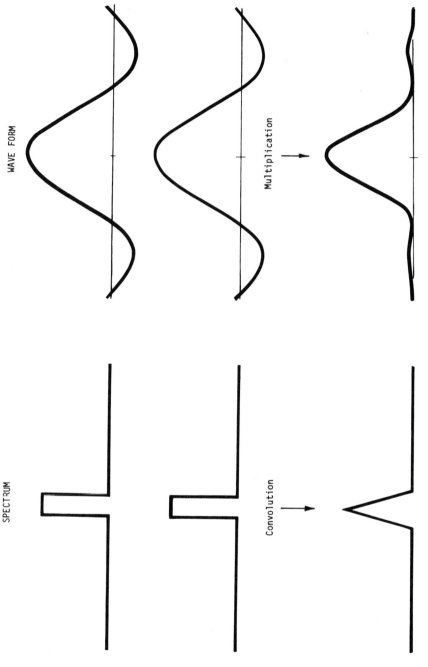

FIG. 9. Self-convolution of a rectangle function to yield a triangle function may alternatively be achieved by multiplication of their FTs.

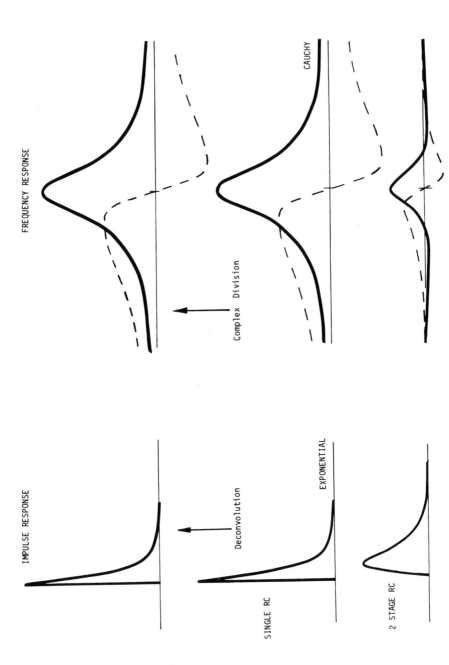

FIG. 10.   Deconvolution by complex division in the Fourier domain.

In the general case of resolution enhancement in spectra, the following therefore applies: if the complex FT of the observed spectrum is divided by the complex transform of the distortion function, then the inverse transform of the complex quotient is indeed the true, undistorted spectrum of the sample. In the limit, it will readily be appreciated that in this method of spectrum sharpening, the amplification of the noise level during the division stage ensures the observance of the oft-stated principle of the conservation of total information. These errors naturally become more serious as the width of the distortion function approaches that of the observed absorption bands, and therefore dictate the limit of usefulness of the method. Jansson and co-workers [55] have pointed out that one advantage of iterative deconvolution is that noise and spurious oscillations increase gradually. As a result they can be controlled at any stage by smoothing. In the Fourier method they may appear immediately after a single operation.

C. Determination of Optical Constants

It is only relatively recently that the determination of optical constants in the mid-IR (4000 to 400 cm^{-1}) has received much attention. This is no doubt partly on account of experimental difficulties, but also because considerable computation is required to extract the results from the observed data. The various methods used for their determination have already been reviewed [57]. Interest in this field has been considerably stimulated by recent progress in band shape analysis, which we shall discuss later (Sec. V).

1. Methods Based on Simple Reflection

The definitive form of the spectrum is the complex refractive index ($\hat{n}$), whose real and imaginary parts are the (common) refractive index (n) and the absorption index (k), respectively, i.e.,

$$\hat{n} = n + ik \tag{4}$$

The optical reflectivity (R) at any wave number is a function not only of n and k, but also of the angle of incidence (θ). Simon [58] showed that if R was measured at two values of θ, then n and k could be directly determined by the solution of two simultaneous equations. Robinson and Price [59] then pointed out that, provided the reflectivity measurements extended over a complete absorption region, data at a single incidence angle were sufficient. The method takes advantage of the Kramers-Krönig integral transform relationship between the phase (θ_R) and the modulus of the amplitude ($|r| = \sqrt{R}$) of the reflected radiation:

$$\theta_R = -\frac{2\nu_0}{\pi} \int_0^\infty \frac{\ln |r|}{\nu^2 - \nu_0^2} \, d\nu + \pi \tag{5}$$

where ν_0 denotes the band center and $|r|$ is the square root of the observed reflectance. The optical constants n and k are then derived from known expressions which relate each to $|r|$ and θ_R. The computation is simplified under conditions of near normal incidence, an approach followed by Schatz and co-workers [60,61].

With the development of the more sensitive ATR technique (Sec. III.C.2), there was a lull in simple reflection spectroscopy for optical constant determination. This has only recently been revived by Querry and Williams and their co-workers in their accurate and detailed studies of liquid water [62], aqueous solutions [63,64], and liquid ammonia [65]. The recent paper by Hale and Querry [66] also describes a novel means of data acquisition in a study of the optical constants of liquid water throughout the 200-nm to 200-µm wavelength range.

2. *Attenuated Total Reflectance (ATR) Methods*

Simple reflectance methods suffer from an intrinsic lack of sensitivity outside regions of very strong absorption. The situation is considerably improved in the ATR situation, where the radiation beam is reflected internally from the boundary between the liquid sample and a highly refracting window material, such as silver chloride, germanium, or KRS-5. In the absence of sample absorption the beam is reflected totally for incidence below the critical angle, with a sudden decrease in reflected intensity as this angle is exceeded. This transition in the R versus θ diagram becomes less abrupt as absorption by the sample layer increases, so that there is a small range of angles around this critical angle in which the internal reflectance provides a very sensitive measure of the optical constants n and k.

The technique was first proposed by Fahrenfort and Visser [67, 68] and has subsequently been used for optical constant determination by, among others, Crawford and co-workers [13-15] and Irons and Thompson [69]. Several variations of experimental procedure have been suggested. Fahrenfort [67] proposed measuring either R_s at two incidence angles, or R_s and R_p (or their ratio) at one angle (the suffixes s and p denote the perpendicular or parallel plane of polarization, respectively), since for these cases analytical expressions could be used to determine n and k directly. It has been more usual, however [13-15,69], to measure instead the unpolarized reflectivity at several angles, and preferably also using several prism materials [69], and to solve sets of simultaneous equations for n and k after the manner of Simon's original graphical method [58]. For the best results, it is essential that great care is taken in the choice of the two incidence angles used.

One of the penalties of the high sensitivity of which the ATR method is capable is that the rather specific optimum angles must be accurately and precisely maintained during the experiment (to less than 10' arc). Moreover, in cases of very strong absorption (where $k_{max} > 0.5$) the critical angle itself can change significantly and several different angle pairs may be required to cover the whole absorption region, which can sometimes result in apparent discontinuities of the resultant n and k spectra. It is of course important that the incident beam striking the reflection surface is properly collimated, so that there is no contribution from angles of incidence other than that specified. Fahrenfort [67] and Crawford and co-workers [13-15] used a hemicylindrical prism as the reflecting optic. So long as the incident beam is focused at the prescribed

distance in front of it, good collimation within is assured. Unfortunately, this adjustment is critical and must be made each time
the optic is changed. The problem of residual beam divergence was
recognized (as the "angle spread effect" [13]) and attempts were
made to correct for it in the computation procedure [70]. These
problems were largely avoided by Irons and Thompson, who used instead a prism with an isosceles triangular section and external beam
collimation. The angle of incidence could then be fixed to within
6' of arc [69].

In view of the considerable experimental difficulties involved,
it is encouraging that there is such good general agreement not only
between these two sets of ATR results but also between them and direct transmission studies [4,69] on the same compounds [Sec. III.C.3].

3. *Thin Film Transmission Techniques*

The principal experimental difficulty of using the direct transmission method for determining the optical constants of strongly absorbing liquids is associated with the preparation and accurate
thickness measurement of the very thin films required (0.5 to 10 μm).
Even more serious, however, is the need to make careful corrections
for the dispersion distortion effects [57] arising from optical interference within the layers of the sample cell, reflection losses
at the outer window surfaces, and anomalous dispersion effects at
the window/liquid interfaces. The magnitude of these effects was
first shown by Yasumi [71] in measurements on carbon tetrachloride,
and subsequently by Maeda and co-workers [72-74] and Fujiyama and
co-workers [75] who have computed the distortion that is likely to
arise in a number of experimental situations.

In the presence of such effects, the apparent absorption index
(k_a) is related to the observed transmittance ($T = I/I_0$) by the
simple expression

$$k_a = \ln{(T^{-1})}/4\pi\nu\ell \tag{6}$$

where ν is wave number (cm^{-1}) and ℓ the film thickness [57,75].
There is unfortunately no explicit expression relating the apparent

and true absorption indices. It was nonetheless recognized by Maeda
and co-workers [73] that, by analogy with Eq. (5), the phase shift
on transmission (θ_T) is related by a Kramers-Krönig transform to the
modulus of the transmitted amplitude ($|t| = \sqrt{T}$):

$$\theta_T = -\frac{2\nu_0}{\pi} \int_0^\infty \frac{\ln|t|}{\nu^2 - \nu_0^2} \, d\nu + 2\pi\nu_0\ell \qquad (7)$$

Therefore, if the transmittance is measured over the whole absorption
region, the integral can be evaluated at each point and the spectrum
of θ_T determined. The optical constants may then be calculated from
two simultaneous equations relating them to t and θ_T. This approach
has been followed by Kozima and co-workers [76] and Neufeld and
Andermann [77] for the case of thin solid films supported on a transparent substrate, with iteration to obtain the most self-consistent
results. Carlon [78] has described a simple method for determining
n only from conventional transmittance spectra for various polymer
films.

In the development of a general method for determining the
optical constants of liquids by transmission through very thin cells
[79], Young and Jones outlined the use of an alternative Kramers-
Krönig transform relating n and k [57]:

$$n = \frac{2}{\pi} \int_0^\infty \frac{\nu k}{\nu^2 - \nu_0^2} \, d\nu + \bar{n} \qquad (8)$$

where $\bar{n}$ is a mean refractive index term arising from contributions
outside the actual range of integration. In practice this can usually be estimated by extrapolation of the wings of the refractive index
spectrum, or alternatively by measuring the interference fringe spacing in the wings using a thicker cell. The iteration procedure has
been described [4] and is similar to the pseudodeconvolution method
of correcting for finite slit distortion [45]. An apparent refractive index (n_a) curve is derived from the measured k_a using Eq. (8),
and these indices are used to compute a doubly distorted absorption

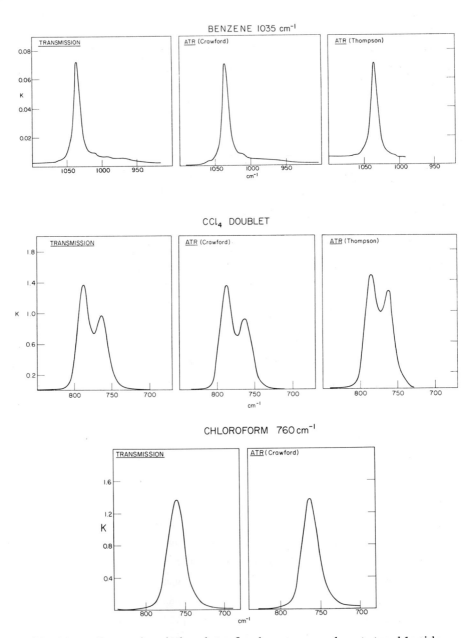

FIG. 11a.   Absorption index data for benzene, carbon tetrachloride, and chloroform obtained by transmission and ATR [14,69] methods. (Figure 11 reprinted from Jones et al. [4], by courtesy of the editor of *Journal of Molecular Structure*.)   See text (p. 398).

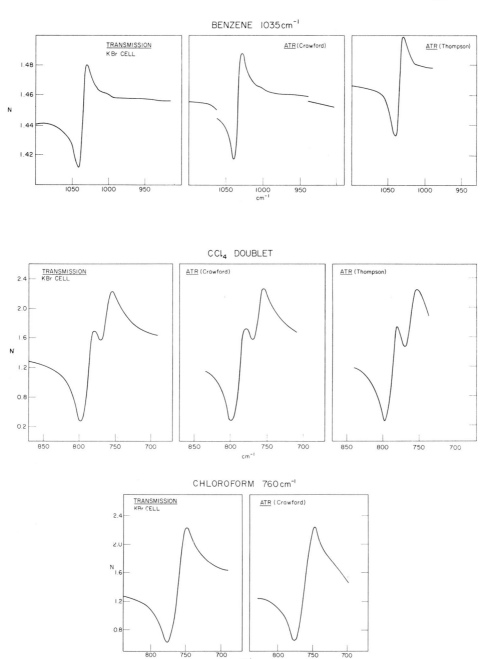

FIG. 11b. Refractive index data for substances in Fig. 11a.

curve (k_{aa}). A first approximation to k is obtained from the difference (k_a - k_{aa}) together with a new n from Eq. (8). Iteration is
continued (about four cycles are usually adequate) until a set of n,
k data is reached, from which the computed k_a curve matches the recorded data within the original noise level. Allowance is made for
beam convergence, nonparallelism of the sample film, and instrumental
polarization. In a preliminary check of the optical equation system,
the computational procedure [80] has proved highly successful in
correctly predicting the reflection loss and interference effects
within empty cells [81]. The computed and observed fringes agree to
within 0.5% T. The transmittance values of single, polished windows
agree to within 0.2% T [4].

Preliminary studies of the absorption spectra of benzene, chloroform, and carbon tetrachloride [4] show a most encouraging agreement
with the published ATR data [14,69], as shown in Fig. 11. Once the
remaining problems of film thickness measurement and the estimation
of $\bar{n}$ in Eq. (8) have been fully resolved, the transmission approach
has the considerable advantage that no modifications need be made to
the spectrometer as are required in the ATR technique. The latter
does, however, have the theoretical advantage that the Kramers-Krönig
transform is avoided, and n and k can both be determined by two angle
measurements at a single wavelength.

We should mention in closing that, although dispersion distortion
can indeed cause serious errors in some circumstances [75], the magnitude of these effects is expected to fall well below the present
limits of measurement precision (about 0.4% T) in all cases where the
cell pathlength exceeds about 20 µm or when using dilute solutions
(<1% molar) [82]. Only if computer-corrected ratio-recording techniques allow an ordinate precision of the order of 0.1% T or better
is it likely that the adequacy of conventional "matched cell" precautions of general purpose quantitative analysis need be questioned.

IV. METHODS OF QUANTITATIVE ANALYSIS

A. Heights and Areas

1. Peak and Valley Location

The program described by Jones [12] using algorithms originally
developed by Savitzky [83] is typical of a number of programs that
have been used for this purpose. Absorption maxima are approximately
located by the change of sign of the first derivative, and the region
around the peak is then fitted with a cubic function according to the
least squares criterion. This equation is differentiated to give a
quadratic whose root defines the peak position and which may then be
used to interpolate the true peak height from the least squares cubic
[17].

The situation is merely inverted if absorption minima are required. Special precautions may have to be incorporated to detect
weak inflection points. In this case the initial first derivative
may not actually change sign, although it will show a discontinuity
that should be recognizable numerically.

In quantitative chemical analysis with carefully measured control runs, a simple but accurate peak height determination may well
be adequate for the purpose. In some cases, the coordinates of the
peak and its neighboring minima may be desired. The accurate determination of cell pathlength is facilitated by the numerical analysis
of fringe patterns using precise peak and valley location [12].

2. Baseline Problems

One of the greatest single problems of applying IR spectroscopy
to quantitative analysis is the difficulty of accurately placing the
baseline. Where this arises because of solvent or impurity absorption, or from general loss of transmission through the sample cell,
the problem can usually be avoided by an appropriate compensation
technique. This is considered in Sec. IV.B. We shall deal briefly
here with the residual question of how to locate the base of a peak
whose accurate height is required when the absorption of neighboring
bands clearly has to be reckoned with.

The simplest approach computationally is to locate the adjacent minima on each side of the band and linearly interpolate the ordinate at the previously determined peak position [84]. Alternatively, one may construct a tangent across the neighboring valleys, which although so easy to do manually with a ruler is not entirely straightforward algorithmically.

Digital techniques naturally lend themselves to a higher degree of sophistication than this. In particular, data that are clearly beyond the wings of the band in question could be specified as baseline regions and a least-squares curve-fitting technique used to join them into a continuous quadratic, cubic, Cauchy, or other profile (see also Sec. IV.C.3). The interpolated ordinate at the peak position again gives the estimated base level for the subsequent measurement. The introduction of a nonlinear baseline is quite legitimate over short wave number ranges in studies of well-defined bands or when compensation is difficult to achieve (either physically or by subtraction, e.g., with pressed disks). It seems less justified (or necessary) over extended ranges of complex spectra, in the case of band fitting, where regions of unexpected absorption are often better catered for by the inclusion of weak extra bands. Perhaps for this reason the NRCC band-fitting programs [18,19] allow only a variable base level although their modification is not difficult. For band shape analysis [19] involving any numerical integration, provision for at least a sloping linear base line over the range of the band is essential.

3. Integrated Areas

There are basically two ways of computing band areas. In the general case one proceeds by numerical integration between predetermined limits. If the data are sufficiently closely spaced, the simple addition of ordinate elements will give a result accurate enough for most purposes. Otherwise, or for very accurate measurements, some form of quadrature should be used, for example, by Simpson's rule or with Newton-Cotes weighting. A simple Fortran program (subroutine Quad) is given in the NRCC bulletins [17-19].

The alternative approach requires that the nature of the band
profile be known or can be safely assumed. Depending on whether a
Cauchy, Gauss, or other function is used, the total area may be ex-
plicitly determined from the characteristic parameters of the basic
function. This technique is useful when the full extent of an absorption band is either inaccessible or overlapped by neighboring
ones. In practically all cases, it will first be necessary to optimize the characteristic parameters by an iterative least-squares
procedure (Sec. IV.C.3) using an appropriate model function. The
area expressions for several commonly used functions [17-19,82] are
given in Table 1.

B. Difference and Derivative Spectra

1. Background Elimination

In many cases, unwanted background absorption can be eliminated
during the recording of a spectrum using a suitable means of compensation in the reference beam of a double-beam spectrometer. For example, a liquid cell containing solvent is used to balance the equivalent absorption in the sample cell; to a first approximation, the
energy lost in various sampling accessories (e.g., microsampling)
may be compensated by a grid or comb attenuator. Such procedures
presuppose good optical balance between the two beams, and optical-
null instruments suffer the added disadvantage of insensitivity under
low-energy conditions. Although the latter problem is largely overcome by ratio-recording photometry, optical-null instruments can also
give accurate and reproducible results through digital data logging,
often in single-beam mode.

The success of such digital compensation techniques depends
very much on the use of an appropriate reference assembly and on the
photometric and environmental stability of the instrument over the
time of data acquisition. The correct reference is usually evident
on a "before-after," "solvent-solution," or "empty-full" basis. For
pure liquids, however, the situation is often more difficult because
neither a single window nor an empty cell accurately represents the

TABLE 1

Area Expressions for Various Absorption Profiles

Type	Ordinate expression[a]	Area expression
Cauchy	$Y = a/(1 + bx^2)$	$A = a\pi/b^{1/2}$
Gauss	$Y = a \exp(-dx^2)$	$A = a(\pi/d)^{1/2}$
Cauchy-Gauss sum[b]	$Y = a/(1 + bx^2) + c \exp(-dx^2)$	$A = a\pi/b^{1/2} + c(\pi/d)^{1/2}$
Cauchy-Gauss product[b]	$Y = a \exp(-dx^2)/(1 + bx^2)$	$A = (a\pi/b^{1/2})\exp(d/b)\mathrm{erfc}(d/b)^{1/2}$
Variable exponent[c]	$Y = a/[1 + (2^{e^2} - 1)bx^2]^{1/e^2}$	$A = a/b^{1/2}[\pi/(2^{e^2} - 1)]^{1/2}\dfrac{\Gamma(1/e^2 - 1/2)}{\Gamma(1/e^2)}$

[a]In all expressions: a and c are peak heights; b and d are width coefficients such that for a Cauchy component $b = 4/(\Delta\nu_{1/2})^2$ and for a Gaussian $d = 4 \ln 2/(\Delta\nu_{1/2})^2$; $x = \nu - \nu_0$, where ν_0 is the peak position (cm^{-1}); $\Delta\nu_{1/2}$ is the half-height width; e is a variable component.
[b]See Ref. 18.
[c]See Ref. 85.

"no-absorption" background [4]. Ideally one should use a nonabsorb-
ing liquid of the same mean refractive index. Since this is in
general not available, the alternative approach is to compute the
liquid cell background transmission from first principles and sub-
tract this from the absorption spectrum actually recorded [4]. The
same technique could be used to remove interference fringe base
lines from the transmission spectra of thin polymer films, provided
that appropriate refractive index values are obtainable and that
beam and sample geometry are known.

Whatever particular method is adopted, there can be no doubt
that digital background elimination is so much more convenient than
manual replotting in situations where it is necessary at all, that
this alone can often justify the installation of a simple data logging
means quite apart from the higher accuracy that is attainable.

   2. Addition and Subtraction of Spectra

Computationally, the subtraction of a complex spectrum is no
different from the elimination of a more featureless background.
The distinction is made partly for the sake of emphasis and also to
indicate the scope of both addition and subtraction in simulation
studies. The quantitative determination of one component in a well-
defined mixture is considerably simplified if the complicating absorptions of other materials can be removed. The residual spectrum
can then be directly quantified by peak height or total area calculation without resort to band-fitting methods that are computationally time-consuming (i.e., expensive) and in which the symmetric
profile assumptions generally introduce small additional errors.

This in itself points to a separate application of spectrum
subtraction, namely, for the comparison of an experimentally observed
spectrum with one computed from it. Pitha and Jones [86], for example, have shown the separation of noise and misfit errors in the
representation of part of a complex steroid spectrum. The degree of
distortion introduced by some process on real or synthetic data is
often best illustrated by the simple difference curve.

Certain features of such results should be recognized lest artifacts be misinterpreted. Quite commonly the difference curve has the appearance of a first derivative (Sec. IV.B.3). This will inevitably result from a misalignment of the two spectra and will affect all bands to some extent, with bands of similar intensity and width being affected equally. If this is recognized, it can be overcome by a retrospective calibration of one set or simply by an arbitrary relative adjustment to bring them into best synchronism. As quite often happens, and for which special allowance must be made, some bands are shifted relative to the others (e.g., solvent effects) and only some parts of the difference curve appear in the derivative form. When this problem arises in quantitative chemical analyses, it can often be avoided by the preparation and proper use of suitable control solutions or mixtures.

It is also possible for difference spectra to appear as a second derivative (Sec. IV.B.3), which usually indicates that the two spectra involved are no longer at the same spectral resolution. This might be because the slit settings were different during data acquisition or because one has subsequently suffered some degree of distortion, perhaps by oversmoothing. It is not unknown, of course, for solvent effects also to give rise to band broadening, which would show a similar result. We recall the reverse of this general phenomenon in the method of resolution enhancement by the addition of the second derivative to the observed spectrum (Sec. III.B.3).

In both qualitative and quantitative analysis, the facility of synthesizing mixture spectra from digital reference data can be most valuable. For example, does the spectrum of this product and so much of that additive actually match the spectrum of the competitor's material? In simulation studies of this kind the trial-and-error addition of stored spectra is much quicker computationally than the alternative band-fitting approach, although it is naturally liable to give erroneous results in cases where solvent and other distortion effects are significant.

Programs for addition and subtraction of spectra, which are
versatile in input-output scale possibilities, have been described
by Jones [12]. Similar features are built into the software package
available with the Digilab FTS-14 interferometer [32].

3. Derivative Spectroscopy

First or second derivative recording offers several advantages
of presentation, provided that the original noise level is not too
high. The principal utility of first and higher derivatives is the
way in which subsidiary features are made to appear more distinct.
This is always at the expense of a progressive increase in noise
level, in accordance with total information conservation.

Although derivative spectra may be computed from simple first,
second, or higher order differences [26], it is more usual and convenient to use the convolution procedure described by Savitzky and
Golay [43]. Because the peak value of the nth order derivative is
inversely proportional to the nth power of the original half-width
(see below), it is especially important to select a convolution
function of appropriate width relative to the bands in the spectrum,
and to recognize that bands of different width will have different
relative peak heights in the various derivative modes. This is
intuitively reasonable, since the first derivative measures the slope
of absorption contours and the second derivative their curvature,
etc. They are not solely related to the original peak height.

It has already been mentioned (Sec. IV.A.1) that the first derivative is useful in locating the positions of absorption maxima.
If the noise level permits, the second derivative shows the same
thing even more clearly. In Fig. 12, for example, the components of
a quintuplet are well revealed by the second derivative [87]. Furthermore, their individual spacings are seen to be virtually identical, a conclusion which could hardly be drawn with confidence from
the original spectrum. This therefore points to one of the two major
applications of derivative spectroscopy, namely, as an accurate peak

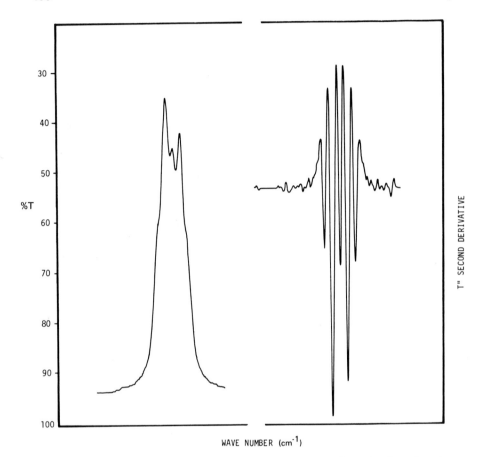

FIG. 12.   Normal and second derivative spectra of a simulated quintuplet.

locator.   To this end, it may profitably be used as a preliminary to
complex band-fitting programs where a good initial estimate of peak
positions is required.   Grum and co-workers [88] have discussed applications in u.v. spectroscopy and presented a number of illustra-
tions of first and second derivative spectra.

The quantitative relationship between a spectrum and its derivatives should not be overlooked, since this is the basis of real
applications in quantitative analysis.   In the general case of an
overlapped mixture spectrum the derivative plot gives, if anything,
a more misleading impression of relative concentrations, since band
widths are usually not accurately known.   However, in a well-defined

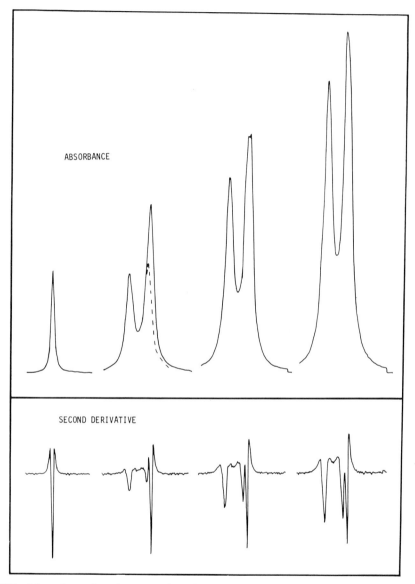

ABSORBANCE

SECOND DERIVATIVE

FIG. 13.   Quantitative application of second derivative recording.
See text (p. 408).

mixture, where the concentration of a single component only is variable and must be determined, derivative data do provide a consistent
and accurate quantitative measure of the amount of that component.

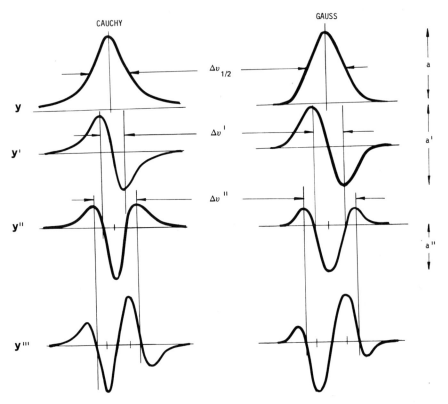

FIG. 14. Cauchy and Gauss profiles, and the first, second, and third derivatives (y, y', y", and y'"). Expressions for the widths ($\Delta v'$ and $\Delta v''$) and amplitudes (a' and a") are given in Table 2.

This would seem to be useful in process control of impurity monitoring, where significant (or even severe) overlap with other bands makes direct peak height measurement impossible. Such a situation is simulated in Fig. 13, where a constant amount of one component (sharp singlet) is mixed with an increasing amount of another (underlying doublet), to the point that the singlet is almost completely obscured to a visual observation [87]. The second derivative peak, on the other hand, remains constant.

For the Cauchy and Gauss model functions (Sec. IV.C.3), the first, second, and third derivatives and their principal characteristics are shown in Fig. 14. Their expressions are listed in Table 2.

TABLE 2

Derivative Functions of Cauchy and Gauss Models[a]

Characteristic		Cauchy[b]	Gauss[c]
Ordinate (Spectrum)	$: y$	$a/(1 + bx^2)$	$a \exp(-bx^2)$
Ordinate (1st deriv.)	$: y'$	$-2abx/(1 + bx^2)^2$	$-2abx \exp(-bx^2)$
Ordinate (2nd deriv.)	$: y''$	$2ab(3bx^2 - 1)/(1 + bx^2)^3$	$2ab(2bx^2 - 1)\exp(-bx^2)$
Ordinate (3rd deriv.)	$: y'''$	$-18ab^2x(bx^2 - 1)/(1 + bx^2)^4$	$-4ab^2x(2bx^2 - 3)\exp(-bx^2)$
Max-min (1st deriv.)	$: a'$	$9a/(2\sqrt{3}\,\Delta\nu_{1/2})$	$4a(2 \ln 2/e)^{1/2}/\Delta\nu_{1/2}$
Negative peak (2nd deriv.)	$: a''$	$-8a/(\Delta\nu_{1/2})^2$	$(-8 \ln 2)a/(\Delta\nu_{1/2})^2$
Max-min separation (1st deriv.) : $\Delta\nu'$ i.e., $y'' = 0$		$\Delta\nu_{1/2}/\sqrt{3}$	$\Delta\nu_{1/2}/(2 \ln 2)^{1/2}$
Maxima separation (2nd deriv.) : $\Delta\nu''$ i.e., $y''' = 0$		$\Delta\nu_{1/2}$	$\Delta\nu_{1/2}[3/(2 \ln 2)]^{1/2}$

[a] The partial derivatives of these and other common band shape functions have been given by Jones and co-workers [18,19] and Fraser and Suzuki [85]. See also Fig. 14.

[b] For Cauchy functions, the coefficient b is related to the half-height width $(\Delta\nu_{1/2})$ by $b = 4/(\Delta\nu_{1/2})^2$; a is the peak height; $x = \nu - \nu_0$, each in wave number (cm^{-1}).

[c] For Gauss functions, $b = 4 \ln 2/(\Delta\nu_{1/2})^2$; a is the peak height; $x = \nu - \nu_0$.

C. Multicomponent Mixtures

1. *Linear Combinations of Known Components*

Quantitative mixture analysis by IR is often said to be beset
with problems. It is true that compared with GC and u.v. the noise
level may be higher, and compared with integrating nmr the analysis
may be more difficult. It is also true that each individual spectrum
has a complicated set of bands that leads to considerable overlap in
mixtures, and that solvent and sampling conditions can sometimes
cause significant changes in the various component spectra. For the
analysis of mixtures of gases and volatile liquids, the pseudo matrix
isolation technique [89] has shown considerable promise, since the
spectral simplification in the low-temperature condensed phase allows
peak heights to be used in many cases.

Of more general application, though, are the various numerical
techniques that are possible with computer assistance via digital
data logging. In this section we shall assume that all the constituents of the mixture (gas, liquid, or solid) are known qualitatively,
and that calibrated reference spectra of each are available. It is
also important that Beer's law holds over the concentration ranges
involved, and that any wave number regions where solvent effects are
serious can be avoided. The situation is illustrated diagrammatically
in Fig. 15, where the mixture Z is composed of unknown amounts (x_1,
x_2, and x_3) of compounds A, B, and C whose spectra are given.

In principle, ordinate measurements at three different wave
number positions would allow the setting up and solving of three
simultaneous equations that would yield x_1, x_2, and x_3 directly. In
practice, however, limited photometric accuracy and high band overlap
combine to make such a simple approach generally inadequate. A very
considerable improvement can be achieved when a large number of data
points is used, i.e., in the overdetermined case using the whole (or
a substantial part) of the spectra concerned.

The spectra in Fig. 15 may be regarded as arrays of ordinate
values denoted by a_{i1}, a_{i2}, a_{i3}, z_i for A, B, C, and Z, respectively,
where i is a running subscript referring to the wave number position.

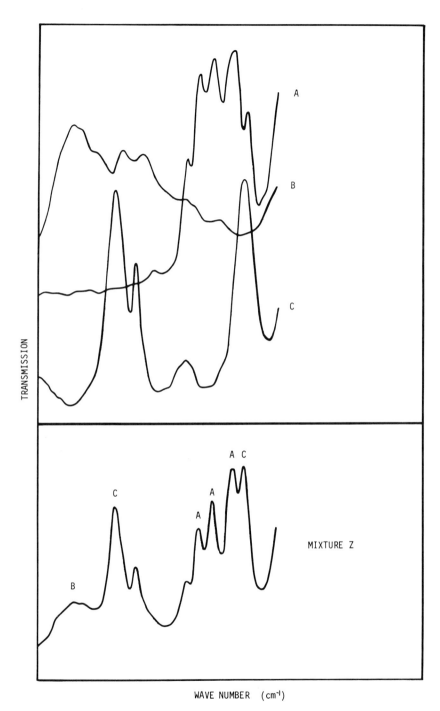

FIG. 15. Spectrum of mixture Z as a linear combination of components A, B, and C.

At each position there will be some error ε_i between values of the observed mixture and the sum of the components. We may therefore express the mixture data in terms of (9) as a linear combination of its components:

$$a_{i1}x_1 + a_{i2}x_2 + a_{i3}x_3 + \varepsilon_i = z_i \tag{9}$$

In the overdetermined case of >3, we shall set up an equation system that satisfies the minimum least squares criterion. From (9) we have:

$$\varepsilon_i^2 = \left[z_i - (a_{i1}x_1 + a_{i2}x_2 + a_{i3}x_3)\right]^2 \tag{10}$$

from which we can write the partial derivatives with respect to each variable x_1, x_2, and x_3:

$$\frac{\partial(\varepsilon_i)^2}{\partial x_1} = 2\left[z_i - (a_{i1}x_1 + a_{i2}x_2 + a_{i3}x_3)\right](-a_{i1}) \tag{11a}$$

$$\frac{\partial(\varepsilon_i)^2}{\partial x_2} = 2\left[z_i - (a_{i1}x_1 + a_{i2}x_2 + a_{i3}x_3)\right](-a_{i2}) \tag{11b}$$

$$\frac{\partial(\varepsilon_i)^2}{\partial x_3} = 2\left[z_i - (a_{i1}x_1 + a_{i2}x_2 + a_{i3}x_3)\right](-a_{i3}) \tag{11c}$$

For a least squares fit, we must minimize the sum of the squares of the error functions:

$$\sum_i^{\nu} \frac{\partial(\varepsilon_i)^2}{\partial x_1} = \sum_i^{\nu} \frac{\partial(\varepsilon_i)^2}{\partial x_2} = \sum_i^{\nu} \frac{\partial(\varepsilon_i)^2}{\partial x_3} = 0 \tag{12}$$

Summing over all i (i.e., over the wave number range), and rewriting (11) for this condition, we have:

$$\sum_i^{\nu} (a_{i1}a_{i1})x_1 + \sum_i^{\nu} (a_{i1}a_{i2})x_2 + \sum_i^{\nu} (a_{i1}a_{i3})x_3 = \sum_i^{\nu} (a_{i1}z_i) \tag{13a}$$

$$\sum_i^\nu (a_{i2}a_{i1})x_1 + \sum_i^\nu (a_{i2}a_{i2})x_2 + \sum_i^\nu (a_{i2}a_{i3})x_2 = \sum_i^\nu (a_{i2}z_i) \qquad (13b)$$

$$\sum_i^\nu (a_{i3}a_{i1})x_1 + \sum_i^\nu (a_{i3}a_{i2})x_2 + \sum_i^\nu (a_{i3}a_{i3})x_3 = \sum_i^\nu (a_{i3}z_i) \qquad (13c)$$

This set of linear simultaneous equations in x_1, x_2, and x_3 can be
solved by any of the standard procedures, but advantage may be taken
of the diagonal symmetry of the matrix of coefficients and the simple
Gauss-Jordan elimination method used by Pitha and Jones [18] can be
followed. It should be noted that there is a unique result for each
variable; no iteration is involved. The method may be extended, in
principle, to any number of components, although the accuracy of the
measurements and the required precision of the calculations will impose some limitation in practice. It should also be borne in mind
that absorbing impurities are bound to interfere to some extent, and
correct results cannot be guaranteed unless all major components are
identified beforehand. Nevertheless, with well-behaved systems and
using appropriate standards, quantitative analyses to better than 1%
should be possible routinely.

2. *Self-modelling Resolution of Principal Components*

It is by no means likely that all the components of all mixtures
submitted for analysis are actually known or that their individual
spectra are always obtainable. Lawton and Sylvestre [90] tackled the
following problem: given the spectra of a series of mixtures known
to contain the same materials in various proportions, how can their
individual spectra be determined and the analysis made quantitative?
They developed a technique based on principal component analysis,
which reduces the mixture data set to a linear combination of eigen-
functions. By applying certain restrictions, they showed that it is
possible to define the form, within a small tolerance band, of the
underlying component spectra [90]. Because no assumptions were made
about the shapes of the bands (e.g., Lorentzian, Gaussian, etc.),
their resolution was described as self-modelling. Even with very

heavy band overlap, the method works well. However, it has only yet
been proven for a small number of components and with data of high
S/N (e.g., u.v./visible spectra).

The above mentioned restrictions in the analysis, which would
somewhat detract from its general applicability in the IR region,
can be eliminated if some model is available for relating the unknown
amounts to some experimental variable such as time or initial concentration [91]. The method would therefore seem to be particularly
useful in studies of chemical equilibria and kinetics and in situations where the various absorbing species cannot be examined in isolation.

3. Band-Fitting Techniques (I)

Quite clearly, there are many situations in the analysis of
quantitative mixtures that are not capable of solution by either the
linear combination or self-modelling methods; e.g., if not all the
constituents are identified, or if band shifts or shape changes in-
validate the assumption of simple additivity, or if no kinetic or
other data are available. Yet it could still be important to quantify
spectral changes, such as relative concentrations based on the integrated areas of overlapped groups of bands. As always, it is necessary to make some kind of assumption before we can proceed. We are
reduced (some would say) to accepting mathematical, functions to describe the individual absorption bands. A number of different profiles have been suggested, using three, four, or five parameters to
represent peak position, peak height, and half-width terms [85,86].
The merits and physicochemical implications of these various types
will be discussed later (Sec. V). In that section we shall mention
some of the band-fitting algorithms that may be used to optimize the
total parameter set.

At first sight the problem seems formidable. Complex spectra
typically contain 20 to 30 absorption bands, each requiring 3 to 5
parameters to be adjusted, and therefore involving the simultaneous
optimization of maybe 100 or more altogether in accordance with a
suitable criterion of "best fit." The best match between the computed

and observed spectra is generally taken to be when the sum of the
squares of the differences is minimized or falls below a threshold
determined by the original measurement uncertainty. Although it is
the absorbance curve which has a profile conforming to the model
function, it is usual to minimize the transmittance differences
during the computation, since it is on this scale that the photometric uncertainty is approximately constant. Different results are
to be expected if absorbance differences are minimized instead, and
the more so the higher the actual absorbance values.

The various optimization methods that have been described fall
into two broad categories: (1) "direct search" methods, which only
require the evaluation of the minimization function itself for many
trial parameter values, and (2) "gradient" methods, wherein the
first derivatives of the function (with respect to each variable) are
also computed to guide the search for the optimum values down the
path of steepest descent.

The simplex method developed by Nelder and Mead [92] from that
of Spendley and co-workers [93] is probably the simplest of all, and
the easiest to program. Although evidently successful [92] with
small numbers of variables, the simplex method has been found to be
progressively less efficient, compared with other methods, as the
number of parameters is increased [94]. Of all the methods compared
by Box [94], the most successful direct search method was that described by Powell [95]. The advantage of also computing the derivatives (if they can be formulated for the system under examination)
was shown [94] by the superior performance of the method originally
proposed by Davidon [96] and refined by Fletcher and Powell [97].
Pitha and Jones made a detailed comparison of a number of gradient
techniques, including the Davidon-Fletcher-Powell method [97] and
concluded that, at least in the application to spectral parameter
optimization, the damped least squares algorithm developed by
Marquardt [98] and Meiron [99] was considerably better than all others
tested. This has formed the basis of the various programs described
by Jones and co-workers [18,19,86]. They differ only in the actual

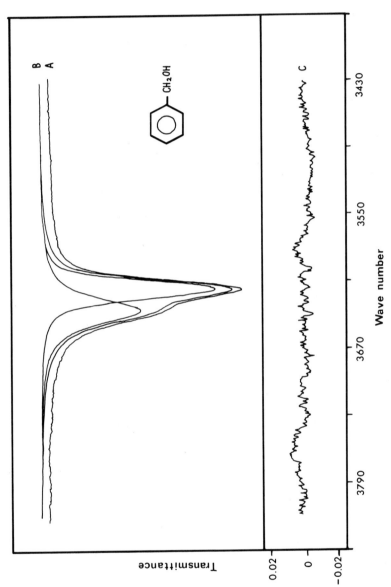

FIG. 16. Part of the IR spectrum of benzyl alcohol in CCl$_4$ solution (A), resolved into two separate Cauchy-Gauss components (B) with the combined noise/misfit error curve (C). (Reprinted from Krueger and Hawkins [100], by courtesy of the editor of *Canadian Journal of Chemistry*.)

function used to represent the absorption profiles. For a more detailed description and the Fortran listings, the reader is referred
to the NRCC bulletins [18,19].

Once the peak heights and widths have been determined for the
various component bands, their relative areas may immediately be
calculated (Sec. IV.A.3, Table 1). Krueger and Hawkins [100] have
recently determined conformational enthalpy differences in substituted benzyl alcohols from changes in computed band areas as a function of temperature, using band fitting to separate the overlapping
absorptions of the two conformers. Figure 16 shows their separation
of the O-H stretching bands of the two conformers of benzyl alcohol
(dilute solution in CCl_4) using two Cauchy-Gauss sum bands convoluted
with a triangular function (equivalent to 5-cm^{-1} slit width). The
high quality of the fit is confirmed by the difference spectrum below. Nowhere is the misfit error greater than 1% T. Several examples of the general applications of band fitting in spectroscopy are
shown in the article by Fraser and Suzuki [85].

V. BAND SHAPE ANALYSIS

A. Method of Truncated Moments

In 1963 Jones and co-workers [101] described a statistical approach to the analysis of IR band profiles which quantified the
shape of a band independently of its intensity or half-width. The
general moments (μ_r) of any absorbance function [$A(\nu)$] about its
peak position (ν_0) are given by:

$$\mu_r = \frac{\int_{-\infty}^{+\infty} (\nu - \nu_0)^r A(\nu) d\nu}{\int_{-\infty}^{+\infty} A(\nu) d\nu} \tag{14}$$

A Gaussian function is characterized by a finite value for each even
moment whereas a Cauchy function, whose wings are indefinitely extended, has all its even moments infinite (except μ_0). These general

moment values are therefore of little use in comparing actual spectra
with model profiles, particularly since individual bands cover only
a limited wave number range. It is nevertheless of interest to plot
the course of the moments (particularly μ_2) as the range of integration is extended out from the band center. The truncated moments
$\mu_r(j)$ are computed from (15):

$$\mu_r(j) = \frac{1}{(\frac{1}{2}\Delta\nu_{1/2})^r} \frac{\int_{-j}^{+j} (\nu - \nu_0)^r A(\nu) d\nu}{\int_{-j}^{+j} A(\nu) d\nu} \tag{15}$$

at increasing j values [where $j = (\nu - \nu_0)/(\frac{1}{2}\Delta\nu_{1/2})$]. Using the
$(1/\frac{1}{2}\Delta\nu_{1/2})^r$ factor reduces the result to a dimensionless quantity
independent of actual band width and therefore characteristic of
only the shape of the function $A(\nu)$. It has been shown [101] that
the second moment (μ_2) for the Gaussian reaches its limiting value
by $j \simeq 3.2$ whereas for the Cauchy profile μ_2 increases linearly
above $j \simeq 1.5$.

    This, therefore, provides a rather useful way of characterizing
band shape from a purely statistical point of view. For example,
the 813-cm^{-1} band of perylene (solution in CS_2) is seen to depart
from the pure Cauchy shape in a well-defined manner when observed
with increasing spectral slit width [101]. It is as valid to use
these numerical methods to elucidate the nature of recognized dis-
torting influences as it is to measure the shape of the fundamental
spectrum itself.

    Such studies have helped to confirm that the character of most
IR bands (of liquids and solutions) is in some way intermediate between the Gauss and Cauchy models and have supported the development
of band-fitting techniques using sum and product combinations (this
is considered in more detail in Sec. V.B). It has, moreover, been
suggested [57] that in suitable cases (i.e., for relatively isolated
bands) analysis of the truncated moment characteristics could prove
to be a useful diagnostic aid in selecting an appropriate function

for band fitting. Kakimoto and Fujiyama [102] have recently reported
the second moment curves of several of the bands of liquid methyl
cyanide. The a_1-type fundamentals are close to Cauchy shape, whereas
the e-type bands have a significant Gaussian component.

It is an inherent property of all symmetric band functions that
the odd moments are all zero. This makes the third moment (μ_3) particularly useful, not only for the detection of asymmetry (which can
be done more simply by other means) but also in its accurate quantification [57,101,103].

As in all attempts to study and measure band shapes in detail,
the importance of eliminating all base line errors cannot be over
emphasized. This applies equally to integral methods, such as truncated moment analysis and dipole correlation functions (Sec. V.C),
as well as to iterative band-fitting techniques. It will be readily
appreciated, for example, that even within a respectably low noise
level, a Gaussian band standing on a small positive base line is
very hard to distinguish from a nearly truly Cauchy function. Similarly, a negative base line connecting the wings of a truncated
Cauchy curve can appear very like a Gaussian.

B. Band Fitting (II)

There is a sound theoretical basis for supposing that IR absorption bands might have a Cauchy (Lorentzian) profile. This model
assumes that the primary factor causing line broadening is disruptive
collisions between molecules leading to random reorientation. This
was recognized by Lorentz [104] and developed by Van Vleck and
Weisskopf [105] for the spectra of simple gases and has since been
generally assumed to apply (to a first approximation) to liquids
also. However, a number of factors are known to cause a departure
from this ideal behavior, some having a chemical origin, others
arising from measurement distortion.

In cases where the splitting due to mixed isotopic species is
smaller than the individual half-widths, the overall band appears to
have some degree of Gaussian character. Similar effects are likely

to result from conformational isomerism. In either case, if the
relative intensities of the components differ widely, the resulting
band envelope may become quite asymmetric. On the other hand, a
band that truly has a Cauchy profile may become distorted during recording by convolution with the slit transfer function, which will
introduce some form of Gaussian perturbation. The effects of these
and other instrumental factors on band shapes have been discussed at
length by Seshadri and Jones [103], the indications being that most
real bands will be best represented by some kind of combination of
these two models (i.e., Cauchy and Gauss).

Pitha and Jones [86] proposed the sum and product combinations;
Voigt [106] a convolution function; Fraser and Suzuki [107] a variable exponent form of the Cauchy profile; and Young [108] a general
polynomial function. All have been used successfully in some situations.

According to Crawford and co-workers [109-111], the corrected
absorption index data obtained by the ATR method for some bands of
liquid methyl iodide and hexafluorobenzene are indeed completely
Cauchy, within the limits of experimental uncertainty. Rosenberg
[112] studied the spectra of aliphatic esters and simple aromatic
hydrocarbons and found that the observed bands were well represented
by pure Cauchy profiles, convoluted with a triangular slit function.
Pitha and Jones [86] found that the Cauchy-Gauss product function
(Sec. IV.A.3, Table 1) was successful in fitting part of a steroid
spectrum, although the five-parameter sum function was generally
preferable [113]. A restricted version of the sum function, where
the Cauchy and Gauss half-width terms were kept constant and approximately equal, has been used by Fraser and Suzuki [85] in matching
the polarized spectrum of silk fibroin, and also by Krueger and
Hawkins [100] in their studies of substituted benzyl alcohols. The
apparent absorption index curve of the 760-cm^{-1} band of a thin film
of liquid chloroform (1.47 μm in KBr) was found to be very well represented [108] by a limited polynomial of the form:

$$k_a = \frac{a}{1 + bx^2 + cx^4} \tag{16}$$

where a = peak height, $x = \nu - \nu_0$ (cm^{-1}), b and c = width coefficients. In conjunction with (16), a simple method was described [108] for estimating the maximum necessary order of the polynomial, based on the limiting slope of the log k versus log $(\nu - \nu_0)$ plot.

It should be clear from the above that at the present time no single band profile can be recommended as being universlaly applicable. The approach has to be an empirical one, using the various methods outlined to aid the selection of an appropriate function.

Only when a large number of measurements has been reported with careful attention paid to the accuracy and precision of the data will any useful generalizations emerge. It seems inevitable that progress in this direction will closely parallel advances in the Fourier-related field of dipole correlation functions, which we shall examine next.

C. Dipole Correlation Functions

Ever since the early work of Lorentz [104] it was recognized that the broadening of liquid state absorption bands was the consequence of the rotations and collisions of the absorbing molecules. It is only rather recently, through the theoretical studies of Gordon [114] and Shimizu [115], that the formal relationship between spectral band shape (IR and Raman) and molecular motion has been made clear. In particular, these studies have shown that the FT of the absorption profile $A(\nu)$ yields a time distribution $\phi(t)$ that directly represents the nature and rate of the average dipole reorientation. Taking only the real part of the transform for symmetrical bands, we have:

$$\phi(t) = \frac{\int_{band} A(\nu) \cos [2\pi c (\nu - \nu_0)t] \, d\nu}{\int_{band} A(\nu) \, d\nu} \tag{17}$$

Two distinct processes have been identified: brief periods of in-
ertial (free) rotation characterized by a Gaussian $\phi(t)$ curve, and
random disruptive collisions occurring over rather long periods with
an exponential decay. As might be expected, the results observed
experimentally usually show a mixed or intermediate behavior but have
been widely interpreted in terms of these two models [57,116]. It
is usual to display log $\phi(t)$ versus time, because the exponential
decay then appears as a linear function whose slope is directly related to the rate of collisional disorientation.

Correlation function data from IR spectra have been reported
for trans-dichloroethylene [117], cyclohexane [118], methyl iodide
[109,111], benzene and hexafluorobenzene [15,110], methane [119],
acetonitrile [102,120], and carbon disulfide [121]. Raman data have
also been used in studies of liquid methane [114], nitrogen, and
oxygen [122].

The precise interpretation of dipole correlation functions is
by no means always straightforward, even though in principle one is
freed from the restrictive assumptions of Cauchy or Gauss spectral
band shape. In practice, however, the separation of the log ϕ curve
into a linear long-time region after an initial nonlinear section is
an implicit acceptance of the same assumptions. Thus, the best
straight line is "fitted" to the long-time data and its point of
departure from the intitial curve is taken as the average period of
essentially free rotation. Close examination of the computed $\phi(t)$
data points in the literature will show that this approach seems
well justified in some cases but hardly at all in others. The problem, of course, is how to deduce the physical implications of doing
otherwise.

At this point, it is instructive to inquire what errors might
be introduced through numerical misadventure. It is well known to
habitués of Fourier analysis [50] that the transform of any spectrum
containing a rectangular component will show the oscillations characteristic of the sin x/x function. We therefore expect to find
such features in any dipole correlation curves obtained from spectra

which are either truncated in their abscissa range or displaced
relative to their true base line. These effects were simulated for
a Cauchy band by Fujiyama and Crawford [111], who found that even
when the integration range of Eq. (17) was over eight times the
width at half-height, oscillations were still observed. Moreover,
in the log $\phi(t)$ plot they are exaggerated and the best fit straight
line can become significantly misplaced. A base line shift of only
1% of the peak value was also shown to cause similar problems. It
is in a sense unfortunate, therefore, that most absorption bands are
in fact predominantly Cauchy, since this is the function that is
most susceptible to truncation and baseline errors in Fourier trans-
formation. Not only is it hard to make accurate allowance for
neighboring bands, and hence avoid truncation, but also careful compensation techniques must be used in all solution studies and very
thin pure liquid samples may require correction for dispersion distortion effects (Sec. III.C.3). The ATR technique, on the other
hand, avoids the baseline problem and should be accurate for very
strong bands, but has other difficulties of its own (Sec. III.C.2).

The limitations of the Gaussian/exponential model having been
recognized, a more general theory of the dynamics of simple liquids
and the implications for IR spectra was recently proposed by Bratoz
and co-workers [123], in which complex expressions were derived for
the absorption profile associated with various relaxation mechanisms.
In the absence of translational diffusion, the Gaussian/exponential
model is confirmed as a representation of simple free rotation plus
collisional diffusion. It was also shown that asymmetric bands could
be handled and indeed should be regarded as the general case [123].
The FT then contains a nonzero imaginary component, and the dipole
correlation function can no longer be computed from Eq. (17) alone,
but must include the sine transform as well [50].

It has already been pointed out [57,124] that, as the Gauss/
exponential model for $\phi(t)$ is the Fourier equivalent of the Cauchy-
Gauss sum function for $A(\nu)$, the same results might alternatively
be obtained by a band-fitting approach. In recent studies of the

fundamental bands of acetonitrile, Kakimoto and Fujiyama [102] and
Yarwood [120] have each compared the results obtained with the two
methods. Both reports confirm the feasibility of fitting either the
spectrum or the correlation function, but the differences between
the two sets of results demand caution in their detailed interpretation. They show an interesting discrepancy: Kakimoto and Fujiyama
[102] find in $A(\nu)$ a narrow Gauss component of comparable or greater
peak height than the Cauchy, whereas Yarwood [120] finds instead a
broader, very weak Gauss component underneath a predominantly Cauchy
profile.

Even when all reasonable steps are taken to reduce experimental
errors, we still seem to be faced with the possibility of alternative
solutions in least squares fitting procedures, whether in the time
or frequency domain. A clear distinction between similar competing
solutions should not be expected, however, until truncation and base
line problems are rigorously eliminated and the measurement precision
is sufficiently increased.

The disagreement between two results using the same model might
alternatively indicate a basic inadequacy of the model to represent
the case in point. We have already seen (Sec. V.B) that other mathematical expressions have been successful in describing IR absorption
bands, although their implications in the time domain have generally
not been emphasized. The Cauchy-Gauss product function, for example,
has proved satisfactory on occasions where the normally preferred
sum function has failed. The associated dipole correlation function
is in this case given by the convolution of the separate Gauss and
exponential transforms. Likewise the limited polynomial Eq. (16),
when factored, can be regarded as the product of two separate Cauchy
functions. The transform of this is the convolution of the two
Fourier-related exponentials. It has already been pointed out [108]
that this particular $\phi(t)$ function can be expressed as the difference
of two exponentials, with the interesting possibility that the overall
relaxation situation might be regarded as two simultaneous random
processes occurring at different rates, analogous to a two-stage

radioactive decay. As the number of exponentials that are convoluted together increases, the result becomes increasingly Gaussian [50]. Likewise in the frequency domain, if the order of the polynomial Eq. (16), or the value of the exponent in Fraser and Suzuki's function (Sec. IV.A.3, Table 1) is raised, then the absorption profile also becomes increasingly Gaussian [107,108].

In conclusion, we should emphasize that IR and Raman band shape analysis is still very much an experimental field. This is especially true in the sense that a considerable amount of accurate experimental work still needs to be done to properly test and consolidate the impressive theoretical advances that have already been made. To this end, computer assistance for both data logging and sophisticated analysis programming will no doubt continue to prove essential.

VI. SPECTRUM STORAGE AND RETRIEVAL

A. Archival Records

The organization and proper maintenance of a collection of chart spectra is central to the operation of most analytical laboratories and has been tackled in a variety of ways. Similarly, there are no rules as such for the best way to keep digital spectra. One very soon discovers how few spectra it takes to overfill the main memory of any laboratory computer, especially if it is necessary to store all the detail contained in extended or closely spaced point-by-point records.

We have already seen (Sec. II.C.2), for example, that a single full-range spectrum may easily contain over 5,000 data points. Even if only the ordinate data are stored and packed two values to a 16-bit word (i.e., 2 x 8-bit bytes with a precision of 0.4%), we shall use over 2,000 memory locations per spectrum. On a time-sharing system this could prove expensive if many spectra have to be kept for long periods; on a laboratory minicomputer system even a medium-size disk unit will soon be filled. There seems little sensible alternative to the unlimited expandability of cartridge disks or tape cassette units in terms of speed of access, convenience, and cost. The

merits and demerits of cards and paper tape have been discussed before (Sec. II.C.2,3). Even the high capacity of a tape cassette
(e.g., 3.6 M bit) should not be abused, however, or problems of in-
dexing and filing the cassettes themselves may become significant.

The reasons for keeping full point-by-point digital records
should be understood at the outset. The most obvious purpose is for
the collection of reference spectra of high quality and accuracy of
new or important compounds which might be made available to other
groups or institutions. A more common requirement might be found in
quantitative analytical applications, where carefully calibrated
standard spectra are frequently used in a particular analysis on the
same spectrometer. Perhaps the widest application, at the present
time, would be simply for comparative visual display purposes using
an oscilloscope, in the same way that one might browse or scrutinize
a collection of chart spectra. If, indeed, all that is required is
an approximate graphical representation to, say, 1 or 2% T, for
qualitative or semiquantitative identifications, it seems wasteful
of storage space to keep full archival records. The alternative,
discussed in the next section, is to store a set of parameter values
(e.g., from band fitting) from which a good approximation to the
original can be regenerated whenever required. In this way an order
of magnitude storage saving can generally be achieved. That ought
not to be ignored.

B. Spectrum Condensates

We have discussed in earlier sections (IV.C.3 and V.B) the
possibilities of fitting IR bands with mathematical functions for
quantitative and theoretical applications. The aspect that was
originally emphasized [1], however, was the degree of data reduction
achieved and its implications for more efficient means of spectrum
storage and retrieval. One can typically expect a storage saving
of at least an order of magnitude. Whereas an archival spectrum
might require 5,000 ordinate values stored in 2,500 words (previous
section), the alternative is to store just the list of band-fitting

parameters (or indices [1]. On the basis of 50 bands each requiring 4 parameters (see Sec. V.B) on a linear sloping baseline, this contains only about 200 numbers instead of the original 2,500. Clearly the actual factor depends on the total number of bands (and hence the length of the parameter list) as against the original data density and wave number range.

The quality of fit that may be achieved through such spectral condensation has been well demonstrated by Jones [1] for part of the spectrum of cyclohexanone (solution in CS_2), as shown in Fig. 17. The regenerated result is clearly quite adequate for practically all comparative purposes. The original 1,540 data points have been reduced to 77 parameters using 19 Cauchy-Gauss product bands plus a horizontal baseline value.

It is only fair to admit that, at the time of writing, no one has actually reported the use of such a system for the automatic condensation and storage of IR data for routine library purposes. The reason is not hard to find. The programs involved are not only complex mathematically (and are therefore written in Fortran, although they are freely available [12,18,19]), but also require a substantial amount of computation time to complete the iteration process. They are therefore too expensive to run on a big machine if the objective is to compile a spectrum library in this way, even if the bulk data transfer problems could be overcome satisfactorily.

As far as laboratory minicomputer systems are concerned, the major obstacle seems to be one of software development. We must learn how to efficiently implement the optimization programs for 100 to 200 variables in a small memory configuration. On the one hand, it is surprising that this has apparently not yet been attempted. Then again, it could be said that the new lost-cost storage media have significantly reduced the motivation.

Nevertheless, with the continuing rapid growth of interest in computer-based spectrum libraries, it still seems important to devise some means of efficient data reduction to a level intermediate between the full point-by-point record and the binary coded search string (see Sec. VI.C), so that at least some degree of intensity and shape information can be preserved.

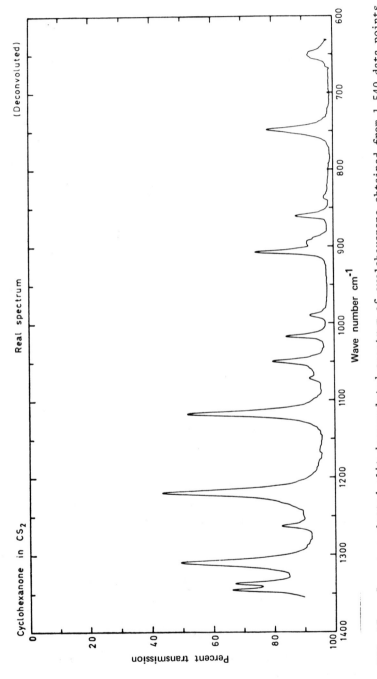

FIG. 17a. Computer-plotted slit-deconvoluted spectrum of cyclohexanone obtained from 1,540 data points encoded at 0.5-cm^{-1} intervals.

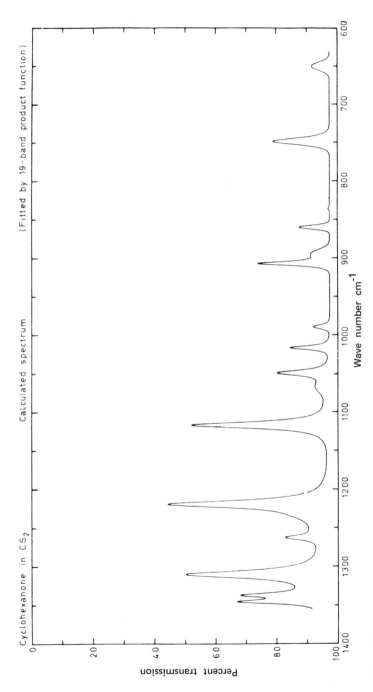

FIG. 17b. Representation of Fig. 17a regenerated from 77 band-fitting parameters. (Reprinted from Jones [1], by courtesy of the International Union of Pure and Applied Chemistry.)

C. Library Search Methods

The archival and condensed spectra we have discussed above represent two approaches to the storage of digital spectra. They can
be retrieved directly by using the name or serial number assigned
by the file organizer. When the name is not known, we are confronted
with a new kind of problem. If the library does contain a spectrum
of the compound, we shall want to find it in order to positively
identify the sample. Alternatively we may know that the library
does not contain that particular spectrum, but the five or ten spectra that most closely match it might help considerably to establish
its chemical type. It is important to make this distinction at the
outset in view of the rather low probability of finding an exact
match even between different spectra of the same compound (solvent,
resolution, concentration, or coding errors), and also because some
search systems cannot find near misses.

In 1967 Anderson and Covert [125] described a computer system
for searching IR spectral data held on magnetic tape. The data base
was compiled from the original ASTM collection of spectral data cards
for over 100,000 compounds. Individual library records are constructed as follows. The spectrum is divided into wavelength intervals of 0.1 μm over the range 2 to 15 μm, and coded according to
whether or not a band exists in each interval, with provision for
declaring any band in each 1-μm region to be particularly strong.
Chemical group and melting and boiling point information are also
included. In any search system, the matching algorithm must operate
within the limitations of the data base and the spectrum of the unknown sample must first be converted into a compatible format. In
the system devised by Anderson and Covert, the sample spectrum was
coded manually onto a special request form, from which the actual
input data cards were prepared [125]. Up to 20 intervals could be
specified for each search, with bands or no bands, and as mandatory
or desirable, together with 15 chemical classification terms. From
this information, the first task of the program is to construct a
search profile for comparison with the library records. These are

then scanned sequentially, the requested features examined for hits
or misses, and a list maintained of the best matches. If any mandatory term is not met, the comparison is terminated immediately to
save time. A band position tolerance of ±0.1 μm is incorporated to
allow for shifts due to solvent effects or uncertainties of coding
bands near interval boundaries (in the sample or the library record).
The final printout consists of a printed listing of serial number
and hit statistics of the 10 (or 100) library spectra that most
closely match the sample.

Erley [126] has described a similar system using an IBM 1130
computer with disk storage and more sophisticated programming which
will scan the same ASTM-based library at a rate of about 10,000 spec-
tra/sec. One feature of the system is that the spectroscopist can
exercise direct control over the search procedure by typing the input data directly into the computer via a TTY. After inspecting the
results which are printed almost immediately (about 90 sec for 90,000
compounds), he is free to modify the search profile as seems appropriate and try again. On receipt of the input data, the program
constructs a binary-coded search string of the sample spectrum in
the exact image of the library records. The actual comparison procedure may then be performed by logical masking operations which
reduce considerably the overall computation time. The position
tolerance check (±0.1 μm) can be made simply by shifting the bits
in the string one place to the left or right.

Erley's original pattern-matching algorithm, SIRCH-I [126], was
superseded first by SIRCH-III [127] with the introduction of a variable tolerance allowing one or more band mismatches, and then by
FIRST-1 [128] which incorporated several improvements. Exact matches
were rated higher than those which only matched after the ±0.1-μm
shift, and important bands could be weighted (i.e., as Anderson and
Covert's "mandatory/desirable" designation [125]). Erley has reported
the results of a quantitative evaluation of these three algorithms
[128] which showed that, whereas the early SIRCH-I program could be
used to considerably better effect by an expert than by a novice,

the later FIRST-1 performed about equally well for both. This is
probably an indication that a near-maximum retrieval efficiency has
been reached and is limited only by the accuracy and precision of
the data base itself.

A number of systems now in routine operation are similar to
[129], or based on, Erley's use of the ASTM file. Some of these
systems are commercially available [130-132]. One of the difficulties of compound identification by computer comparison of library
records is the interference caused by absorbing impurities and hence
the need for careful purification beforehand. The rather complicated
problem of multicomponent mixture identification has been tackled by
Sebesta and Johnson [133], although so far with only partial success.

A completely different kind of search technique has been described by Lytle [134] using an inverted file of data obtained from
Sadtler's Spec Finder system (Sadtler Research Laboratories, Inc.,
Philadelphia, Pa. 19104). Instead of having a collection of coded
spectra (for a number of compounds) where each element, or bit position, represents the presence or absence of an absorption band (i.e.,
a normal file), an inverted file can be regarded as a directory of
all the compounds having one of a set of particular characteristics,
analogous to optical-coincidence cards. The Spec Finder system is a
listing for all compounds of the actual wavelength position (to 0.1 μm)
of the strongest band in each 1-μm interval, from 2 to 14 μm. The
distribution of bands (or no bands) in each interval has been plotted
by Lytle [134], from which it is clear that only the 6- to 14-μm
range is likely to be useful for inverted file searching. Suppose
the unknown spectrum has six significant bands. Those particular
subsets of the inverted file are scanned (by simple logical operations directly on the binary computer words) to find which compounds
appear in each one, the total score giving the degree of match with
the unknown.

This method is characterized by a very efficient means of coding
and therefore uses a minimal amount of computer time and storage
space, but has the disadvantage that the addition of new compounds to

the library involves a complete recompilation [134]. An even higher degree of code compression was reported by Lytle and Brazie [135] using a 13-digit base-eleven number system. The retrieval performance for the resulting 45-bit search strings was very good, and further compression to 16 bits enabled processing speeds of 18,000 spectra/sec to be achieved using an XDS Sigma 5 computer with disk storage facilities.

Rann [136] has more recently described a somewhat similar but remarkably simple system, also based on the Spec Finder data. The spectrum is divided into 10 regions (4000 to 3000, 3000 to 2000, and 2000 to 400 cm^{-1} every 200 cm^{-1}). The strongest band in each is assigned a decimal digit according to its relative position within the interval. In this way each spectrum can be represented as a 10-digit number. When comparing sample and reference spectra, the sum of the modulus of the differences for each digit field very simply gives a useful measure of their spectral similarity. The whole process has been implemented on a desk-top calculator (Hewlett-Packard type 9100 B) interfaced to a TTY. It was estimated that with a high-speed reader unit it should be possible to search the coded spectra at a rate of about 1,500 per minute.

Jurs [137] has discussed the use of hash coding to achieve near-optimum computer searching of general information files. Although it is possible with this technique to search extremely rapidly (more than 20,000 spectra/sec), it was pointed out by Lytle [138] that the method only applies to exact matching and cannot retrieve spectra that are similar.

Cross-correlation techniques have recently been used by Horlick [139] in an attempt to quantify the similarity between different spectra having some features in common. It remains to be seen whether the rather more complex computation involved, and the need for greater detail in the library, will be offset by any particular advantages in the application to qualitative IR comparisons.

We should not close without a brief mention of learning machines. Artificial intelligence and machine decision making are fascinating

fields in themselves. As far as applications in chemistry are concerned, they have been more widely applied to MS than to IR spectroscopy. The objective is rather different than comparative library
searching in that it attempts to be truly interpretive. There are
basically two steps involved. Firstly, the computer is provided with
a training set of spectra and the corresponding structural details
(chemical classification). When these have been learned by establishing appropriate weight vectors, the program can then be tested
on an unknown sample, the result of which is a prediction of the most
likely chemical class. Some progress has already been made in this
direction using IR data [140], but it is still too early to say
whether or not this approach, which seems a natural complement to
comparative spectral searching, will ever come to be as widely used.

ACKNOWLEDGMENTS

I wish to express my grateful thanks to Dr. M. A. Ford and
Mr. W. E. Morgan, Perkin-Elmer Ltd., for numerous helpful discussions
and for suggestions that have helped to improve the accuracy of the
manuscript.

REFERENCES

1. R. N. Jones, *Pure Appl. Chem.*, *18*, 303 (1969).

2. H. Günzler, *Chem.-Ing.-Tech.*, *42*, 877 (1970).

3. C. Spector, *Int. Laboratory*, *Mar./Apr.* 1972, p. 48.

4. R. N. Jones, D. Escolar, J. P. Hawranek, P. Neelakantan, and
    R. P. Young, *J. Mol. Struct.*, *19*, 21 (1973). Paper presented
    at International Conference on Molecular Spectroscopy, Wroclaw,
    Poland, 1972.

5. J. Swalen in *Proceedings IBM Scientific Computing Symposium
    on Computers in Chemistry,* IBM Data Processing Division, New
    York, 1969.

6. R. J. Bell, *Introductory Fourier Transform Spectroscopy,*
    Academic Press, New York, 1972.

7. G. Horlick and H. V. Malmstadt, *Anal. Chem.*, *42*, 1361 (1970).

8. G. Horlick, *Anal. Chem.*, *43*, 61A (1971).

9. H. M. Pickett and H. L. Strauss, *Anal. Chem.*, *44*, 265 (1972).

10. P. R. Griffiths, *Anal. Chem.*, *44*, 1909 (1972).

11. J. A. Decker, *Anal. Chem.*, *44*, 127A (1972).

12. R. N. Jones, *Appl. Optics*, *8*, 597 (1969).

13. A. C. Gilby, J. Burr, and B. Crawford, *J. Phys. Chem.*, *70*, 1520 (1966).

14. A. C. Gilby, J. Burr, W. Krueger, and B. Crawford, *J. Phys. Chem.*, *70*, 1525 (1966).

15. A. A. Clifford and B. Crawford, *J. Phys. Chem.*, *70*, 1536 (1966).

16. R. E. Biggers in Report of Second Meeting of Study Group for Computer Aided Experimental Spectroscopy, Schenectady, New York (unpublished), 1967.

17. R. N. Jones, T. E. Bach, H. Fuhrer, V. B. Kartha, J. Pitha, K. S. Seshadri, R. Venkataraghavan, and R. P. Young, *Nat. Res. Council Can. Bull. No. 11* (1968).

18. J. Pitha and R. N. Jones, *Nat. Res. Council Can. Bull. No. 12* (1968).

19. R. N. Jones and R. P. Young, *Nat. Res. Council Can. Bull. No. 13* (1969).

20. D. G. Larsen, *Anal. Chem.*, *45*, 217 (1973).

21. J. S. Mattson and A. C. McBride III, *Anal. Chem.*, *43*, 1139 (1971).

22. R. E. Dessy and J. A. Titus, *Anal. Chem.*, *45*, 124A (1973).

23. G. Lauer and R. A. Osteryoung, *Anal. Chem.*, *40*, 30A (1968).

24. L. Ramaley and G. S. Wilson, *Anal. Chem.*, *42*, 606 (1970).

25. W. Krag, N. Daggett, R. N. Davis and F. Perkins, *Rev. Sci. Instrum.*, *40*, 1606 (1969).

26. V. R. Sakalys, *Int. Laboratory*, *Mar./Apr.* 1972, p. 11.

27. G. K. Klauminzer, *Appl. Optics*, *9*, 2183 (1970).

28. C. N. Reilley in Report of 6th Meeting of Study Group for Computer Aided Experimental Spectroscopy, Chapel Hill, North Carolina (unpublished), 1972.

29. T. Chuang, G. Misko, I. G. Dalla Lana, and D. G. Fisher, in *Computers in Analytical Chemistry*, Plenum Press, New York, 1970.

30. J. R. Scherer and S. Kint, *Appl. Optics*, *9*, 1615 (1970).

31. M. Shapiro and A. Schultz, *Anal. Chem.*, *43*, 398 (1971).

32. *Digilab News*, *Vol. 1*, No. 1, Digilab, Inc., Cambridge, Mass., 1970.

33. S. P. Perone, *Anal. Chem.*, *43*, 1288 (1971).

34. A. C. Poppe in *The Applications of Computer Techniques in Chemical Research*, Inst. Petroleum, London, 1972.

35. J. G. Hallett, P. A. Lawson, D. Richards, and H. M. Stanier in *The Applications of Computer Techniques in Chemical Research,* Inst. Petroleum, London, 1972.

36. J. C. Elkins, *Int. Laboratory, Nov./Dec.* 1972, p. 23.

37. A. M. Deane, C. Kenword, and A. J. Tench, *AERE Report No. R 7020* (1972).

38. C. E. Klopfenstein, *Int. Laboratory, Mar./Apr.* 1973, p. 35.

39. J. E. Stewart, *Appl. Optics, 2,* 1303 (1963).

40. R. N. Jones, *J. Japanese Chem., 21,* 609 (1967).

41. H. W. Thompson, *Pure Appl. Chem., 1,* 537 (1961); republished as *Tables of Wavenumbers for the Calibration of Infrared Spectrometers,* Butterworths, London, 1961.

42. D. H. Anderson and N. B. Woodall in *Physical Methods of Chemistry,* Vol. 1, Pt. 3B (A. Weissberger and B. Rossiter, eds.), J. Wiley and Sons, Inc., New York, 1972.

43. A. Savitzky and M. J. E. Golay, *Anal. Chem., 36,* 1627 (1964).

44. J. Steinier, Y. Termonia, and J. Deltour, *Anal. Chem., 44,* 1906 (1972).

45. R. N. Jones, R. Venkataraghavan, and J. W. Hopkins, *Spectrochim. Acta, 23A,* 925 (1967).

46. R. N. Jones, R. Venkataraghavan, and J. W. Hopkins, *Spectrochim. Acta, 23A,* 941 (1967).

47. J. P. Porchet and Hs. H. Günthard, *J. Physics E* (Sci. Instr.), *3,* 261 (1970).

48. K. Yamashita and S. Minami, *Japan J. Appl. Phys., 10,* 1097 (1971).

49. S. P. Cram, S. N. Chesler, and D. B. Cottrell, Paper 376 presented at Pittsburgh Conference on Analytical Chemistry and Applied Spectroscopy, Cleveland, Ohio, 1972.

50. R. Bracewell, *The Fourier Transform and Its Applications,* McGraw-Hill, New York, 1965.

51. J. F. A. Ormsby in *Spectral Analysis: Methods and Techniques* (J. A. Blackburn, ed.), Marcel Dekker, Inc., New York, 1970.

52. A. Roseler, *Infrared Phys., 6,* 111 (1966).

53. P. A. Jansson, R. H. Hunt, and E. K. Plyler, *J. Opt. Soc. Amer., 58,* 1665 (1968).

54. P. A. Jansson, *J. Opt. Soc. Amer., 60,* 184 (1970).

55. P. A. Jansson, R. H. Hunt, and E. K. Plyler, *J. Opt. Soc. Amer., 60,* 596 (1970).

56. D. Steele and I. R. Hill, unpublished results, 1973.

57. R. P. Young and R. N. Jones, *Chem. Rev.*, *71*, 219 (1971).

58. I. Simon, *J. Opt. Soc. Amer.*, *41*, 336 (1951).

59. T. S. Robinson and W. C. Price, *Proc. Phys. Soc., London, Sect. B*, *66*, 969 (1953).

60. P. N. Schatz, S. Maeda, and K. Kozima, *J. Chem. Phys.*, *38*, 2658 (1963).

61. D. W. Barnes and P. N. Schatz, *J. Chem. Phys.*, *38*, 2662 (1963).

62. A. N. Rusk, D. Williams, and M. R. Querry, *J. Opt. Soc. Amer.*, *61*, 895 (1971).

63. M. R. Querry, R. C. Waring, W. E. Holland, G. M. Hale, and W. Nijm, *J. Opt. Soc. Amer.*, *62*, 849 (1972).

64. G. M. Hale, W. E. Holland, and M. R. Querry, *Appl. Optics*, *12*, 48 (1973).

65. C. W. Robertson and D. Williams, *J. Opt. Soc. Amer.*, *63*, 188 (1973).

66. G. M. Hale and M. R. Querry, *Appl. Optics*, *12*, 555 (1973).

67. J. Fahrenfort, *Spectrochim. Acta*, *17*, 698 (1961).

68. J. Fahrenfort and W. M. Visser, *Spectrochim. Acta*, *18*, 1103 (1962).

69. G. M. Irons and H. W. Thompson, *Proc. Roy. Soc., Ser. A*, *298*, 160 (1967).

70. W. C. Krueger, Ph.D. thesis, University of Minnesota, Minneapolis, 1966.

71. M. Yasumi, *Bull. Chem. Soc. Japan*, *28*, 489 (1955).

72. S. Maeda and P. N. Schatz, *J. Chem. Phys.*, *35*, 1617 (1961).

73. S. Maeda, G. Thyagarajan, and P. N. Schatz, *J. Chem. Phys.*, *39*, 3474 (1963).

74. S. Maeda, *Bull. Tokyo Inst. Techn.*, *No. 59*, 17 (1964).

75. T. Fujiyama, J. Herrin, and B. Crawford, *Appl. Spectrosc.*, *24*, 9 (1970).

76. K. Kozima, W. Suetaka and P. N. Schatz, *J. Opt. Soc. Amer.*, *56*, 181 (1966).

77. J. D. Neufeld and G. Andermann, *J. Opt. Soc. Amer.*, *62*, 1156 (1972).

78. H. R. Carlon, *Appl. Optics*, *8*, 1179 (1969).

79. R. P. Young and R. N. Jones, *Can. J. Chem.*, *47*, 3463 (1969).

80. R. N. Jones, R. P. Young, P. Neelakantan, and D. Escolar, *Nat. Res. Council Can. Bull. No. 15*, in preparation.

81. R. N. Jones, private communication, 1972.

82. J. P. Hawranek, P. Neelakantan, R. P. Young, and R. N. Jones, *Spectrochim. Acta, 32A,* 85 (1976).

83. A. Savitzky, *Anal. Chem., 33,* 25A (1961).

84. E. G. Brame, J. E. Barry, and F. J. Toy, *Anal. Chem., 44,* 2022 (1972).

85. R. D. B. Fraser and E. Suzuki in *Spectral Analysis: Methods and Techniques* (J. A. Blackburn, ed.), Marcel Dekker, Inc., New York, 1970.

86. J. Pitha and R. N. Jones, *Can. J. Chem., 44,* 3031 (1966).

87. R. P. Young, unpublished results, 1971.

88. F. Grum, D. Paine, and L. Zoeller, *Appl. Optics, 11,* 93 (1972).

89. M. M. Rochkind, *Spectrochim. Acta, 27A,* 547 (1971).

90. W. H. Lawton and E. A. Sylvestre, *Technometrics, 13,* 617 (1971).

91. E. A. Sylvestre, W. H. Lawton, and M. S. Maggio, *Technometrics, 16,* 353 (1974).

92. J. A. Nelder and R. Mead, *Computer J., 7,* 308 (1965).

93. W. Spendley, G. R. Hext, and F. R. Himsworth, *Technometrics, 4,* 441 (1962).

94. M. J. Box, *Computer J., 9,* 67 (1966).

95. M. J. D. Powell, *Computer J., 7,* 155 (1964).

96. W. C. Davidon, *AEC Res. and Dev. Report,* ANL-5990, 1959.

97. R. Fletcher and M. J. D. Powell, *Computer J., 6,* 163 (1963).

98. D. W. Marquardt, *J. Soc. Ind. Appl. Math., 11,* 431 (1963).

99. J. Meiron, *J. Opt. Soc. Amer., 55,* 1105 (1965).

100. P. J. Krueger and B. F. Hawkins, *Can. J. Chem., 51,* 3250 (1973).

101. R. N. Jones, K. S. Seshadri, N. B. W. Jonathan, and J. W. Hopkins, *Can. J. Chem., 41,* 750 (1963).

102. M. Kakimoto and T. Fujiyama, *Bull. Chem. Soc. Japan, 45,* 3021 (1972).

103. K. S. Seshadri and R. N. Jones, *Spectrochim. Acta, 19,* 1013 (1963).

104. H. A. Lorentz, *Koninkl. Ned. Akad. Wetenschap. Proc., 8,* 591 (1906).

105. J. H. Van Vleck and V. F. Weisskopf, *Rev. Mod. Phys., 17,* 227 (1945).

106. W. Voigt, *Munch. Ber.,* 603 (1912).

107. R. D. B. Fraser and E. Suzuki, *Anal. Chem., 41,* 37 (1969).

108. R. P. Young, paper presented at International Conference on Computers in Chemical Research and Education, Ljubljana, Yugoslavia, 1973.

109. C. E. Favelukes, A. A. Clifford, and B. Crawford, *J. Phys. Chem.*, *72*, 962 (1968).

110. T. Fujiyama and B. Crawford, *J. Phys. Chem.*, *72*, 2174 (1968).

111. T. Fujiyama and B. Crawford *J. Phys. Chem.*, *73*, 4040 (1969).

112. A. S. Rosenberg, Ph.D. thesis, Oklahoma State University, 1970.

113. J. Pitha and R. N. Jones, *Can. J. Chem.*, *45*, 2347 (1967).

114. R. G. Gordon, *J. Chem. Phys.*, *39*, 2788 (1963); *40*, 1973 (1964); *41*, 1819 (1964); *42*, 3658 (1965); *43*, 1307 (1965).

115. H. Shimizu, *J. Chem. Phys.*, *43*, 2453 (1965); *48*, 2494 (1968).

116. G. E. Ewing, *Accounts Chem. Res.*, *2*, 168 (1969).

117. J. T. Shimozawa and M. K. Wilson, *Spectrochim. Acta*, *22*, 1609 (1966).

118. W. G. Rothschild, *J. Chem. Phys.*, *49*, 2250 (1968).

119. A. Cabana, R. Bardoux, and A. Chamberland, *Can. J. Chem.*, *47*, 2915 (1969).

120. J. Yarwood, *Advan. Molec. Relax. Processes*, *5*, 375 (1973).

121. M. Kakimoto and T. Fujiyama, *Bull. Chem. Soc. Japan*, *45*, 2970 (1972).

122. M. Scotto, *J. Chem. Phys.*, *49*, 5362 (1968).

123. S. Bratoz, J. Rios, and Y. Guissani, *J. Chem. Phys.*, *52*, 439 (1970).

124. R. P. Young in *The Applications of Computer Techniques in Chemical Research*, Inst. Petroleum, London, 1972.

125. D. H. Anderson and G. L. Covert, *Anal. Chem.*, *39*, 1288 (1967).

126. D. S. Erley, *Anal. Chem.*, *40*, 894 (1968).

127. D. S. Erley, paper presented at Pittsburgh Conference on Analytical Chemistry and Applied Spectroscopy, Cleveland, Ohio, 1970.

128. D. S. Erley, *Appl. Spectrosc.*, *25*, 200 (1971).

129. L. H. Cross, J. Haw, and D. J. Shields in *Molecular Spectroscopy*, Inst. Petroleum, London, 1968.

130. SIRCH (IR Spectral Data Retrieval System), ASTM, Philadelphia.

131. IRGO (IR Spectral Search Service), Singer Technical Services, Inc., New York.

132. IRIS (IR Information Service), Sadtler Research Laboratories and University Computing Co., Heyden and Son, Ltd., London.

133. R. W. Sebesta and G. G. Johnson, *Anal. Chem.*, *44*, 260 (1972).

134. F. E. Lytle, *Anal. Chem.*, *42*, 355 (1970).

135. F. E. Lytle and T. L. Brazie, *Anal. Chem.*, *42*, 1532 (1970).

136. C. S. Rann, *Anal. Chem.*, *44*, 1669 (1972).

137. P. C. Jurs, *Anal. Chem.*, *43*, 364 (1971).

138. F. E. Lytle, *Anal. Chem.*, *43*, 1334 (1971).

139. G. Horlick, *Anal. Chem.*, *45*, 319 (1973).

140. B. R. Kowalski, P. C. Jurs, T. L. Isenhour, and C. N. Reilley, *Anal. Chem.*, *41*, 1945 (1969).

Chapter 6

ORGANIC MATERIALS

Richard A. Nyquist

and

R. O. Kagel*

Analytical Laboratories
Dow Chemical Company
Midland, Michigan

I.   INTRODUCTION. 442
II. EXPERIMENTAL. 443
     A. Sample State. 443
     B. Quantitative Analysis 446
     C. Microsampling 449
     D. Infrared Emission Techniques. 456
     E. Miscellaneous Sampling Techniques 461
III. GROUP FREQUENCIES IN MOLECULAR STRUCTURE
     DETERMINATIONS. 463
     A. Theory. 464
     B. Carbonyl Stretching, $\nu_{C=O}$ 464
     C. C-S, S-H, and S-S Groups. 467
     D. Carbon-Carbon Double Bond Stretching, $\nu_{C=C}$. 469
     E. Triple Bond Stretching, $\nu_{C\equiv C}$ and $\nu_{C\equiv N}$ 471
     F. Organophosphorus Compounds. 472
     G. NO_2 and SO_2 Groups. 473
     H. Carbon-Halogen Stretching 474
     I. Substituted Benzenes. 475

*Present address: Environmental Services, Dow Chemical Company, Midland, Michigan.

J. Azines, C=N-N=C Stretching 480
K. Miscellaneous Raman Correlations 482
L. Miscellaneous Infrared Correlation Charts. 485
    Figures for Section III. 488

REFERENCES 558

APPENDIX: INDEXES OF SPECTRA AND CHARTS 560
A. Infrared Spectra--Alphabetical 560
B. Infrared Correlation Charts. 561
C. Infrared Spectra--Empirical Formula. 561
D. Raman Correlation Charts 562
E. Raman Spectra--Alphabetical. 562
F. Raman Spectra--Empirical Formula 563

I. INTRODUCTION

Infrared (IR) and Raman spectroscopy are well known as useful tools
for investigation and determination of chemical structures. Their
greatest utility lies in their unique application to the identification of chemical functional groups from vibrational spectra. In
this chapter, IR and Raman characteristic vibrational group frequency
data are presented for organic compounds with emphasis on the technique for which certain chemical functional groups are most easily
identified. Moreover, it will be shown how the two techniques complement each other in the identification of chemical structures.

We presume that the reader is already familiar with the background theory of vibrational spectroscopy, and accordingly have devoted little space here to this subject; the reader is referred to
appropriate texts [1-4]. While the basic experimental methods of
IR and Raman spectroscopy are briefly summarized, a more detailed
accounting of new and recent developments in this area is presented
in the text.

Special attention is given Raman spectroscopic techniques,
since in recent years there has been a resurgence of activity in
chemical applications of Raman spectroscopy. The fact that a stable
high-energy coherent source of radiation - the laser - is now a
routine part of Raman instrumentation, more than any other single
factor has provided new impetus for continuing rapid growth in this
area of spectroscopy.

Recent developments in Fourier transform (FT) IR spectroscopy, especially microsampling and emission techniques, are also presented.

II. EXPERIMENTAL

A. Sample State

Infrared and Raman spectra can be measured of pure samples or mixtures in the solid, liquid, or gas phases. Gas phase Raman work requires a high-power output laser (normally 1 W) and since most industrial laboratories utilize smaller lasers (250 mW and under), relatively little work has been done in this area. A more general requirement for gas phase work is that the material have a reasonable vapor pressure at room temperature (unless a hot cell is used), and most complex organic materials do not meet this requirement. The solid and liquid phases represent the major proportion of samples encountered in organic IR and Raman spectroscopic work. The standard techniques for preparing these samples will be summarized since they have been discussed in detail by several authors [4-7].

1. Solid Phase

Solid phase samples are prepared for the IR either as mulls or pellets. The choice between these techniques is still a subject of some debate. In general, the mulling technique is preferred. The sample is ground into a fine powder (average particle size < 2.5 μm) and suspended in an oil. The slurry is smeared between two windows. With the appropriate choice of oils, usually nujol (from 4000 to 1333 cm^{-1}) and fluorolube (1333 cm^{-1} to 400 cm^{-1}), a spectrum can be obtained with minimal interferences from the oil. The pellet technique involves mixing the finely ground sample (average particle size 2.5 μm) with KBr (or KI) and compressing the mixture in an appropriate die using a hydraulic press (pressure between 500 and 1000 atm) [5]. The result is a clear transparent pellet which will yield a spectrum free from medium interferences. There is always the possibility of a reaction between the sample and KBr under the conditions used in pelletizing; ion exchange is quite common when dealing with inorganic samples [8]. Many laboratories restrict the

use of KBr pellets to microsamples, since much less sample is required by this technique than by the mulling technique. Finally,
precautions must be taken to keep the KBr dry or the spectrum will
show water bands at 3400 and 1640 cm^{-1}. It is also difficult to
prepare an acceptable KBr pellet if the KBr is wet.

Solid sample preparation for the Raman is minimal and straightforward. A few milligrams of solid is placed in a 1.5-mm O.D. melting point capillary cell and aligned in the sample compartment of
the spectrometer. The 1.5-mm O.D. capillary is a universal Raman
cell, though either smaller- or larger-size capillary cells are
often used depending on the sample. Depolarization ratios cannot be
easily measured on polycrystal solids because of reflection from the
crystal surfaces. In order to measure depolarization ratios of
polycrystal powders, either the substance must be heated to the melt
or dissolved in an appropriate solvent.

Potassium bromide pellets mounted at 45° to the incident laser
beam have also been successfully used as a means of solid sampling
in the Raman. Spectra of single crystals as small as 0.10 μm x
40 μm x 0.5 mm and weighing approximately 200 ng (or 25 ng in the
beam) have been obtained for characterization purposes by mounting
the crystal on a goniometer [9]. Different orientations of the
crystal can also be examined by this technique.

Solid state spectra of the same substance may vary considerably
if the substance can exist in more than one crystal form. For example, Fig. 1 shows spectra of different crystal modifications of
3,5-dinitro-o-toluamide. Such differences can be extremely useful
when identification of a specific crystalline state is necessary.
However, in general, spectral differences due to crystal modifications often complicate a spectrum to a point where a simple identification becomes a major task. For this reason, wherever possible,
solution spectra are preferred over solid state spectra.

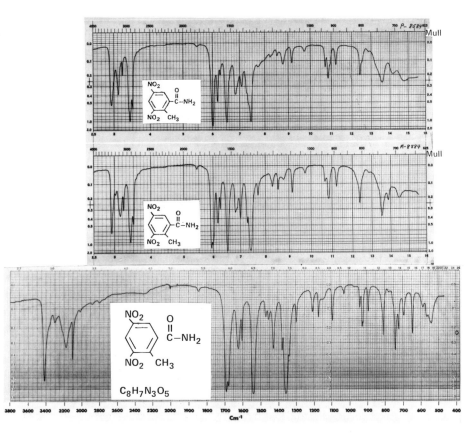

FIG. 1. 3,5-Dinitro-o-toluamide. Upper: IR spectrum; Nujol mull
(2.5 to 15.5 μm). Middle: IR spectrum; Nujol mull (2.5 to 15.5 μm).
Lower: IR spectrum; split mull [Fluorolube (3800 to 1333 cm^{-1}) and
Nujol (1333 to 450 cm^{-1})]. The lower and middle spectra are designated crystalline form I and the upper spectrum is designated crystalline form II.

2. Liquid Phase

Pure liquids with reasonably high boiling points can be studied
by IR as a film between plates or in solution. Volatile liquids are
usually contained in cells where pathlengths between 0.01 and 0.1 mm
are used. The smaller pathlength cells are often difficult to fill

and to clean and for this reason should not be used for viscous substances. Intermolecular effects such as hydrogen bonding are likely
to predominate in the spectrum of a pure liquid run neat. These
effects can be detected and minimized if the compound is dissolved
in a nonpolar solvent and run as a solution. Likewise, other effects
such as rotational isomerism can be verified only if the compound is
studied as a function of concentration in solution or at variable
temperatures [10].

     Whenever possible, it is preferable to prepare the compound as
a solution using nonpolar solvents to minimize solvent-solute interactions. Carbon disulfide and carbon tetrachloride are the most
commonly used IR solvents. A 10% (weight/volume) solution of carbon
tetrachloride scanned from 4000 to 1333 cm^{-1} and a 10% solution of
carbon disulfide scanned from 1333 to 400 cm^{-1} using 0.1-mm NaCl and
KBr cells, respectively, will yield a spectrum relatively free of
interferences due to the solvents. Numerous other solvents have
been used for IR studies but many of these have bands which obscure
large regions of the spectrum. Jones and Sandorfy have tabulated
transmitting regions for a large number of common solvents used in
the IR [11].

     In Raman spectroscopy liquids are normally run neat in a 1.5-mm
O.D. capillary cell. A variety of solvents including carbon tetrachloride and carbon disulfide can be used either singularly or in
combination to minimize solvent band interferences. Water is the
most significant difference between the IR and Raman where the choice
of solvent is concerned. Water is a relatively weak Raman scatterer
and will not affect the pyrex capillary cells.

B. Quantitative Analysis

     Raman can be particularly useful for quantitative analyses in
water solution where IR is restricted because of water band interference. In nonaqueous systems, however, IR is probably the better
method from a precision and accuracy viewpoint. Quantitative IR
spectroscopy is treated in detail in numerous texts [4-7,11]. IR is
also the method of choice for doing trace ppm work but there are some

exceptions to this also. Quantitative analysis by Raman, however, is an area often overlooked by most chemists, and deserves further comment.

The quantitative analysis of acrylonitrile (VCN) in water can be readily obtained by application of Raman spectroscopy. Figure 2 shows the Raman spectrum of a 5% VCN-H_2O solution. Either the C≡N or C=C stretching vibration bands (2240 and 1616 cm^{-1}, respectively) can be used to measure VCN quantitatively. However, there is less chance of interference at 2240 cm^{-1} than at 1616 cm^{-1}, so the 2240-cm^{-1} band is the best choice for this analysis. The first step in performing a quantitative analysis is to prepare a calibration curve using concentrations which bracket the range of concentration in the unknown solution. In Fig. 3, the band peak heights are plotted against concentration which covers the range between 0.1 and 0.5% (1000 and 5000 ppm). The peak height varies linearly with concentration. Each point is the average of four runs. Under normal operating conditions, the scan time is set equal to twice the time it takes to scan over the half-band width of the band. The spectral slit is chosen so that it is less than the half-band width and then the period adjusted so that the pen is alive. For quantitative analysis the instrument is set up in the same way except the scan time is set equal to four times the time it takes to scan over the half-band width.

Once the calibration curve has been constructed, the peak height of the unknown is measured as shown in Fig. 4. This is a typical run at the 0.2 to 0.5% level. It is then a relatively simple matter to read the concentration off of the calibration curve shown in Fig. 3.

The instrument should be relatively stable over a period of a few hours. Instrument stability can be checked by rerunning the standards every 20 to 30 min. However, long-term stability is not good. Changes occur in the laser output and photometry which require a new calibration curve to be constructed on successive days. Fortunately, one set of standards sealed in capillaries can be used almost indefinitely.

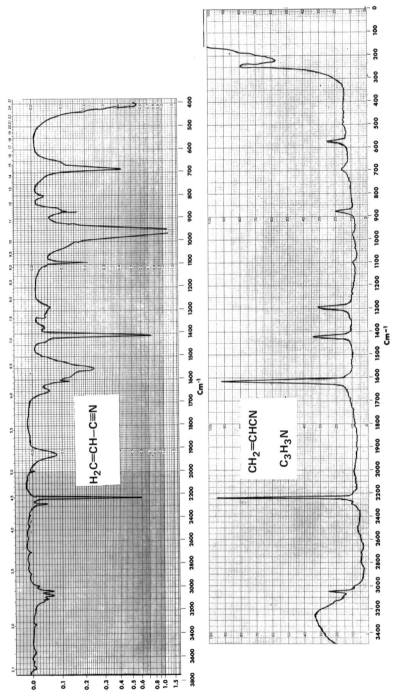

FIG. 2. Acrylonitrile. Upper: IR spectrum; 10% in CCl$_4$ solution (3800 to 1333 cm^{-1}) and 10% in CS$_2$ solution (1333 to 450 cm^{-1}). Lower: Raman spectrum; 5% in water solution. Solvents are not compensated.

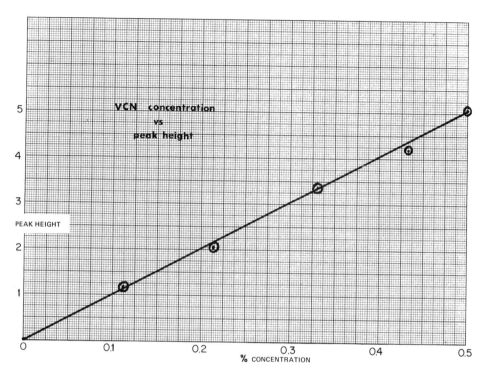

FIG. 3. Calibration curve; percentage VCN versus peak height (x 10⁻¹) of $\nu_{C\equiv N}$, 2240 cm⁻¹. The concentration of acrylonitrile in the unknown solution (Fig. 4) corresponding to a peak height of 37.6 divisions is 0.36%.

In solution, VCN can be quantitated from about 500 ppm up to high percentages. The lower limits of detection and quantitation depend on the laser power: with 1-W laser the lower limit is expected to be 125 ppm.

C. Microsampling

Recent advances in IR Fourier transform spectroscopy (FTS) have generated a new dimension in solid sample handling. Utilizing a 6X Perkin-Elmer beam condenser in the sample compartment of a Digilab Model FTS-14 Spectrometer, spectra have been obtained with as little

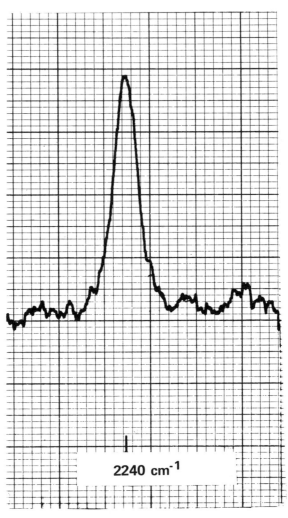

FIG. 4. $\nu_{C\equiv N}$, 2240 cm^{-1}, of a water solution containing an unknown concentration of acrylonitrile. The peak height is 37.6 divisions.

as 20 ng of sample in a KBr pellet. Pieces of material as small as 50 μm in diameter can be conveniently handled by FTS systems. For example, Fig. 5 shows the spectrum of a black speck 0.25 mm in diameter found in a polymer film (upper curve). The lower curve is the

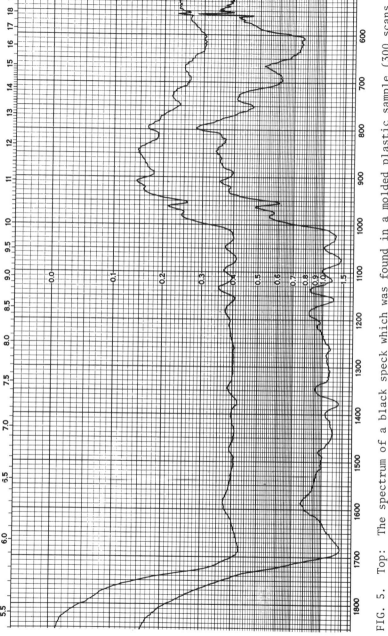

FIG. 5. Top: The spectrum of a black speck which was found in a molded plastic sample (300 scans ratioed against 300 stored scans of the background). Bottom: the spectrum of a portion from a dead insect which was found in the storage bin of granulated polymer feed stock.

spectrum taken directly from a portion of a dead insect found in the
storage bin of the granulated polymer feedstock [12]. This technique
has been extensively applied in conjunction with microscopic techniques to obtain topological spectra of polymers and other materials
[13]. Conventional IR microsampling apparatus can also be used with
most FT systems [14].

One of the greatest assets of the laser as a Raman source is its
compatibility with small samples, especially when the beam is focused
to a spot ∿50 μm in diameter. Microtechniques have been developed
over the recent past which have steadily decreased the sample size
required to produce a usable spectrum from milliliters in the prelaser

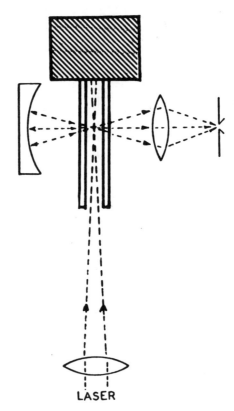

FIG. 6. Axial excitation/transverse viewing geometry. The minimum
sample size with this configuration is about 4 nl.

days to microliters, and presently to less than a nanoliter [15-17].

Figure 6 shows one experimental arrangement for obtaining spectra of small nonvolatile liquid samples. It consists simply of a small 100 μm diameter, open-ended capillary mounted on a goniometer head for ease of alignment. The cell is illuminated axially and the Raman-scattered radiation is viewed transverse to the cell axis. Using 4880 Å argon ion illumination (Coherent Radiation Model 52MG Ar/Kr mixed gas laser), spectra of as little as 4 nl of benzyl alcohol were obtained on a Spex Ramalog system. This sampling arrangement is not particularly suited for work involving volatile materials. The sampling geometry shown in Fig. 7 uses transverse excitation viewing with a sealed capillary cell, the diameter of which can be as

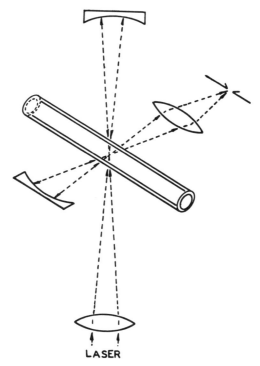

LASER

FIG. 7. Transverse excitation/transverse viewing geometry. The minimum sample size with this configuration is about 0.008 μl [S. K. Freeman and D. O. Landon, *Spex Speaker, 4*, 1 (1969)].

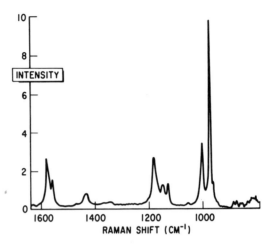

FIG. 8.   Portion of a spectrum obtained with 0.25 nl of benzyl alcohol using the transverse excitation (4880 Å argon ion illumination)/
transverse viewing geometry (5 cm^{-1} resolution).

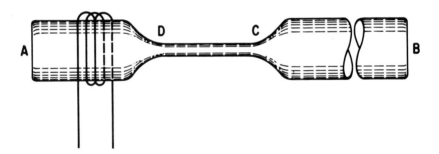

FIG. 9.   Gas chromatographic capillary trapping system.   The center
of a 1.5 mm O.D. capillary is constricted.   End A is attached to the
exit port of a GC unit.   Region D is cooled via a small metal coil,
the ends of which are immersed in liquid nitrogen.   The capillary is
sealed at C and A and the sample centrifuged into the constriction
between C and D.

small as 20 μm.   Our best effort to date with this geometry is shown
in Fig. 8.   This is a portion of the spectrum obtained from 0.25 nl
of benzyl alcohol using 4880 Å argon ion illumination.   Microtechniques are not so much limited by sample size as they are by methods
of handling and transferring the amount of sample needed to produce
a usable spectrum.   This problem has been recognized by a number of

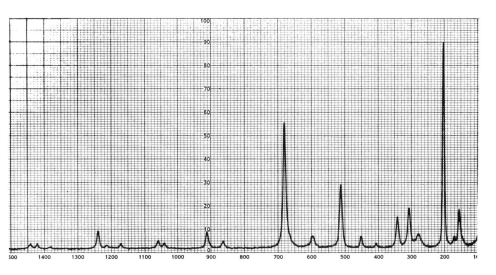

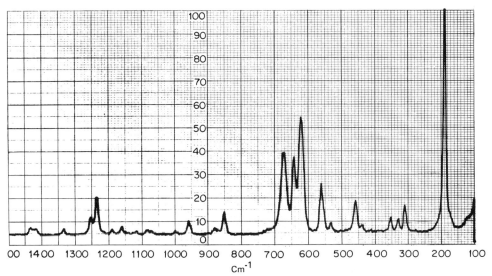

FIG. 10. Raman spectra of trapped GC fractions. Upper: 2 μl of
1,2,2-tribromopropane. Lower: 20 nl of 1,2,3-tribromopropane.
6328 Å He/Ne excitation, 5 cm^{-1} resolution.

people who have taken advantage of the situation by developing gas
chromatography (GC) trapping techniques specifically designed for
Raman applications. Figure 9 shows a microsampling system designed

by the group at Dow 5 years ago which is very similar to that described by Bulkin and co-workers [18]. The center of a 1.5-mm O.D.
capillary is constricted and end A is attached to the exit port of
a GC unit. If the region D is cooled with liquid N_2 using a small
coil, the majority of sample will collect just below the constriction in portion D. The capillary is sealed at C and A and the sample
centrifuged into the constriction then sealed at D. Figure 10 shows
spectra obtained by GC trapping 1,2,2-tribromopropane cuts. The top
spectrum was obtained from 2 µl of 1,2,2-tribromopropane and the
bottom spectrum from 20 nl of 1,2,3-tribromopropane. Both were obtained by using this trapping method. The 1,2,3 isomer represented
less than 1% of the original sample and has escaped detection by IR
analysis. It is felt that one can comfortably handle nanoliter-size
samples using this method.

The obvious advantage of employing a GC trapping technique is
that the spectroscopist usually obtains a sample free from fluorescing impurities which hopefully contains only one component.

D. Infrared Emission Techniques

Fourier Transform Infrared emission spectra of solids, liquids,
and gases have been reported by several authors [19-22]. One basic
experimental arrangement for obtaining emission spectra of solids,
shown in Fig. 11, is quite simple. Solid samples are mounted on a

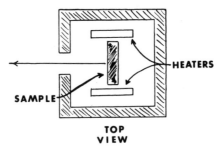

TOP
VIEW

FIG. 11. Basic experimental emission arrangement. The sample is
mounted on a metal stand between vertical heaters in an insulated
"bird-house" enclosure. The aperture is about 1 inch in diameter.

metal stand between vertical heaters and contained in an insulated
"bird house" enclosure. The aperture is about 1 in. in diameter; the
sample temperature is monitored by a calibrated thermistor readable
to about 0.25°C.

Figure 12 shows a series of emission spectra obtained on a
Digilab Model FTS-14 Spectrometer of successively stacked layers of
polyethylene film. All of the spectra were obtained at 60°C and
ratioed against a black body [22]. This experiment was first reported
in the literature by Rhee and Cousins who used Saran* film [21]. They
used it to illustrate one of the major shortcomings of IR emission
spectroscopy, which is the effect of sample thickness on the emission
spectrum. The upper left spectrum is a single layer of polyethylene,
approximately 10 μm in thickness, placed on an aluminum mirror. A
specular mirror is the worst emitter and, therefore, an ideal substrate for this type of experiment. The spectrum clearly shows discrete emission bands of the CH_2 deformations and CH_2 rocking modes.
With 4, 11, and 27 layers these emission bands merge into a broad
band centered around 1300 cm^{-1}. Experiments were repeated using
sheets of known thickness as shown in Fig. 13 to rule out insulation
effects due to trapped air, etc., with the same results. At 0.5 mm
thickness the band structure is no longer discernible and recorded
spectrum resembles that of a "gray body."

The above effect was well known to molten salt chemists.
Kozlowski [23] in 1968 attributed the phenomenon to self-absorption,
i.e., radiation from molecules located in the inner layers is absorbed by those in the outer layers of the sample. The effect is
important if there is a temperature gradient across the sample as is
probably the case here where plastic insulators are being studied.
This points out the need for better sample heating techniques.

The dependence of the discrete emission spectrum on sample
thickness is indeed a limitation on the one hand, but if put into
proper perspective can be used advantageously.

Figure 14 shows the FMIR (frustrated multiple internal reflectance) and emission spectra of a piece of 3/8-in.-thick plastic-

---

*Trademark of The Dow Chemical Company.

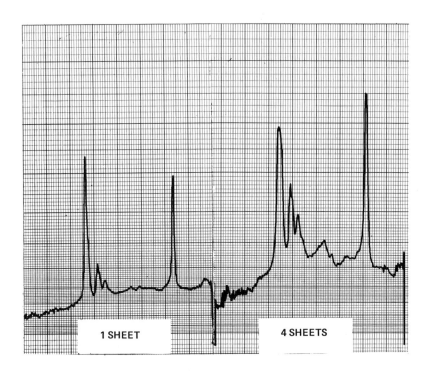

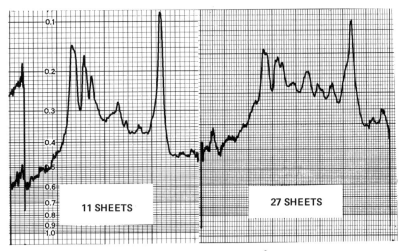

FIG. 12. Emission spectra (2000 to 400 cm^{-1}) of successively stacked layers of polyethylene film. The spectra were obtained with the samples at 60°C and ratioed against a black body at the same temperature.

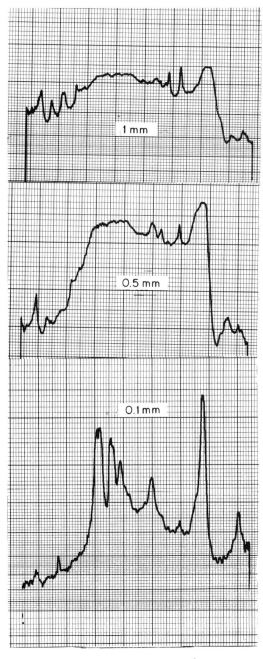

FIG. 13. Emission spectra (2000 to 400 cm^{-1}) of thick single sheets
of polyethylene. The spectra were obtained with the samples at 60°C
and ratioed against a black body at the same temperature.

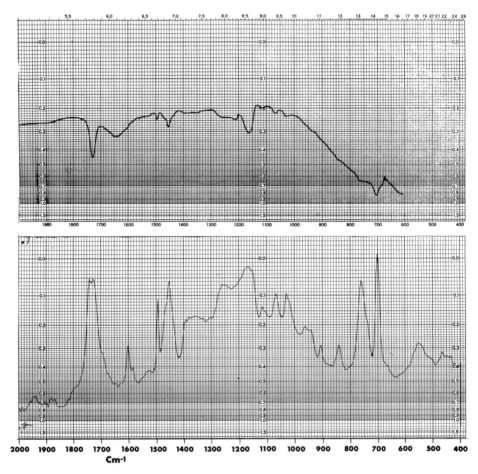

FIG. 14. Comparison between the multiple internal reflection (FMIR) spectrum (top) and emission spectrum (bottom) of a plastic coated piece of aluminum metal.

coated aluminum. The surface of the sample was rough and the sample rigid; hence, poor optical contact with the KRS-5 prism and a poor FMIR spectrum results. The emission spectrum clearly yields more detail. The monosubstituted benzene ring is easily characterized as is the ester carbonyl. Note the reabsorption of self-emitted radiation at 1740 cm^{-1}. The same technique is also quite useful in characterizing thin coats of grease, paint, etc., on metal surfaces [20].

E. Miscellaneous Sampling Techniques

While some Raman sampling techniques such as the GC trapping
are tedious and complicated, this is not the general case. In fact,
one of the beauties of Raman spectroscopy lies in simplicity of
sampling. Occasionally we are forced to deal with an unusual sample
such as that shown in Fig. 15. This is a fuse and it was originally
submitted to GC for analysis of the liquid inside the glass envelope:
this would require breaking the glass envelope to get at the liquid.
The Raman spectrum of the liquid was obtained simply by placing the
entire fuse across the sample compartment and recording its spectrum.
Spectra can be easily obtained of a material in a bottle or ampule
without even opening the container. Once a sample is sealed in a
capillary cell it can be stored indefinitely and an entire standard
collection can be maintained in a standard-size lab bench drawer.

Raman is usually not thought of as the method of choice for
detecting low levels (ppm) of materials in solution as is IR. This
is because of the linear relationship between peak height and concentration. A number of workers, however, have found Raman to be
quite useful in the detection of trace quantities of material in
solution. In the example previously cited, VCN in water solution
has been quantitated at the 500-ppm level and detected at about the
200-ppm level. Strong scatterers such as benzene or any monosubstituted benzene can be detected as low as 10 ppm (Fig. 16). The lower
limit for identification in this case would be about 50 ppm. A number
of materials of this type have been examined at these levels.

The lower limit of detectability decreases proportionally with
increased laser power. For example, the limit of detection for SO_4^{-2}
in water is about 210 ppm using the 4880 Å line of a 250-mW argon
ion laser as the source of excitation and about 50 ppm using the same
line of a 1-W argon ion laser [24]. However, there is a practical
working limit imposed on laser input power, and Brown and Baldwin
[24] estimate this to be about 1 W. Increased laser input power
results in broadening of the Rayleigh line which can extend as far
as 1000 cm^{-1} into the spectrum; this makes the measurement of a band

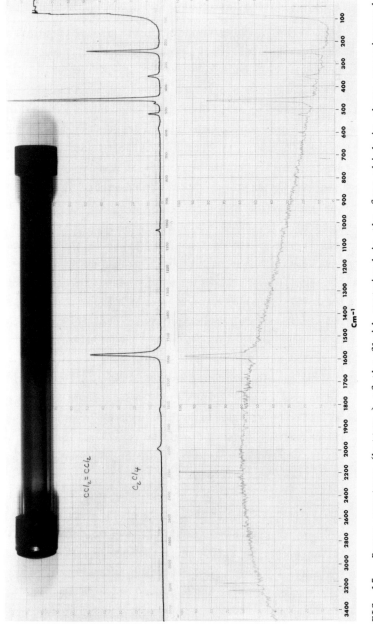

FIG. 15. Raman spectrum (bottom) of the fluid contained in the fuse which is shown superimposed on the studied spectrum of $Cl_2C=CCl_2$ (top). The bands at 2200, 3170, and 3250 cm^{-1} are ghosts.

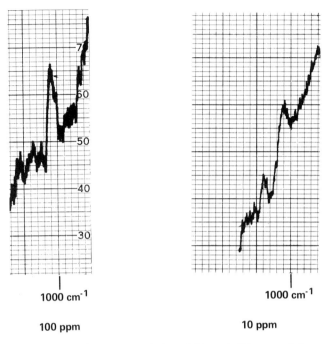

FIG. 16.   Water solutions containing 100 and 10 ppm of benzyl alcohol.

difficult to impossible since one is forced to work on the sharply
decreasing wing of the Rayleigh line.

Raman techniques are usually not sensitive enough to compete
with other methods for trace level detection: the Resonance Raman
effect is an exception to this [25].   In general, it is probably
better to use Raman techniques for examining percent level components
and in this way free those methods which are more sensitive.

III.   GROUP FREQUENCIES IN MOLECULAR
STRUCTURE DETERMINATIONS*

Vibrational spectra are useful in the determination of chemical
functional groups in a molecule because a given vibration occurring
in a functional group tends to have the same frequency and produce

---

*All figures for Section III (Figs. 17-82) appear on pp. 488-
557.

an IR absorption band or a Raman scattering band of the same intensity, to first approximation, independent of the rest of the molecule.
Moreover, small frequency variations for a given vibration of a chemical group in different molecules tend to be systematic and are, in
fact, so systematic that adjacent groups of a class always affect the
group frequency in question in the same manner. Therefore, a study
of group frequency variations allows one to learn more about the adjacent structure in the molecule. These group frequencies have been
discussed in several books [4,7,11,26-28], and it is only our purpose
here to demonstrate the advantages of using both IR and Raman data in
the elucidation of chemical structure.

A. Theory

In order for a vibration to be Raman-active, a change in polarizability must take place during the vibration, while IR requires a
change in the dipole moment. Therefore, molecules with a center of
symmetry will have one set of fundamental vibrations which are only
Raman-active and one set of fundamental vibrations which are only
IR-active. The Raman spectrum emphasizes vibrations centered in
highly polarizable bonds (viz., C-S, S-S, C-I, etc.) while IR emphasizes vibrations centered in polarized bonds (viz., C=O, NO_2, P=O,
SO_2, etc.). In the IR, antisymmetric vibrations usually produce more
intense bands than do the symmetric vibrations while in the Raman the
band intensities are just reversed for antisymmetric and symmetric
vibrations. However, in many cases, IR and Raman spectra yield overlapping data so it is advantageous to compare IR and Raman data whenever possible. Examples are given below.

B. Carbonyl Stretching, $\nu_{C=O}$

Generally, $\nu_{C=O}$ gives relatively strong IR absorption bands
while corresponding Raman bands are relatively weak. The IR and
Raman spectra of diethyl phthalate are compared in Fig. 17. The band
at 1725 cm^{-1} assigned as $\nu_{C=O}$ in the IR has considerably more relative
intensity than does the corresponding Raman band. Since $\nu_{C=O}$ occurs

in a region where few other group frequencies absorb, carbonyl groups
can usually be readily identified from either IR or Raman spectra.
There are cases, though, where more than one IR absorption band occurs
in the region of the spectrum expected for $\nu_{C=O}$, and difficulty can
arise in interpreting these data in terms of useful chemical information.   In these cases, a Raman spectrum can often add additional information.   For example, the IR and Raman spectrum of N-methylphthalimide are compared in Fig. 18.   The IR spectrum shows a strong band
at 1721 cm^{-1} with a medium-intense band at 1759 cm^{-1}.   The band intensities are reversed in the Raman spectrum, and the two sets of
data indicate that two carbonyl groups are present which are joined
to a common nitrogen atom.   The data also show that the in-phase
$\nu_{(C=O)_2}$ band occurs at higher frequency than the out-of-phase $\nu_{(C=O)_2}$
band.   These data are contrary to most symmetric and antisymmetric
vibrational assignments in that the antisymmetric vibration usually
occurs at higher frequency than the symmetric vibration.

The carboxylic acids give very different IR and Raman spectra,
particularly in the $\nu_{C=O}$ region of the spectrum.   The IR and Raman
spectra of acetic acid are compared in Fig. 19.   The Raman spectrum
shows a band at 1674 cm^{-1} which is not apparent in the IR spectrum.
This band results from the in-phase symmetrical $\nu_{(C=O)_2}$ stretching
vibration of the intermolecular hydrogen-bonded dimer (viz.,

).   This dimer has a center of symmetry and
the symmetric $\nu_{(C=O)_2}$ vibration is active in the Raman and inactive
in the IR.   Moreover, the antisymmetric $\nu_{(C=O)_2}$ vibration assigned

at 1710 cm^{-1} (viz.,) is IR-active and

Raman-inactive.   The Raman band at 1764 cm^{-1} results from $\nu_{C=O}$ of

acetic acid monomer. The IR and Raman spectra of benzoic acid are
compared in Fig. 20. The IR band at 1690 cm^{-1} results from antisymmetric $\nu_{(C=O)_2}$ and the Raman band at 1640 cm^{-1} results from symmetric $\nu_{(C=O)_2}$. Both antisymmetric and symmetric $\nu_{(C=O)_2}$ are lower
in frequency than the corresponding vibrations in acetic acid, and
this reflects the effect of conjugation on $\nu_{C=O}$ which weakens the
C=O bond. The IR and Raman spectra of aspirin (acetyl salicyclic
acid) are compared in Fig. 21. The weak Raman band at 1759 cm^{-1} and
the corresponding strong IR band are assigned to $\nu_{C=O}$ of the acetate
group. The IR band at 1701 cm^{-1} is assigned to antisymmetric
$\nu_{(C=O)_2}$ of the $(COOH)_2$ groups, while the Raman band at 1635 cm^{-1} is
assigned to symmetric $\nu_{(C=O)_2}$ of the $(COOH)_2$ groups.

The carboxylic acid salts give rise to antisymmetric and sym-

metric CO_2 stretching vibrations such as $R-C\begin{smallmatrix}O \\ O\end{smallmatrix}^{\ominus}$ and $R-C\begin{smallmatrix}O \\ O\end{smallmatrix}^{\ominus}$

$(\nu_{antisymmetric\ CO_2}$ and $\nu_{symmetric\ CO_2}$, respectively). The IR and
Raman spectra of sodium acetate are compared in Fig. 22. The strong
band at 1583 cm^{-1} and the medium-intense band at 1445 cm^{-1} in the IR
spectrum are assigned as $\nu_{antisymmetric\ CO_2}$ and $\nu_{symmetric\ CO_2}$, respectively. In the Raman spectrum, only $\nu_{symmetric\ CO_2}$ is observed.

The IR and Raman spectra of benzoyl chloride are compared in
Fig. 23. In the IR, the bands at 1737 and 1777 cm^{-1} are assigned to
the Fermi resonance doublet of $\nu_{C=O}$ and the first overtone of the
870-cm^{-1} fundamental. In the Raman, the corresponding bands are
weak but have approximately the same band intensity ratio as observed
in the IR spectrum. It must be noted that the strong IR band at
870 cm^{-1} is not apparent in the Raman spectrum; all benzoyl chlorides
give a strong IR absorption band in the region 900 to 1000 cm^{-1}.

The IR and Raman spectra of benzaldehyde are compared in Fig.
24. The Raman band at 1701 cm^{-1} for $\nu_{C=O}$ is comparatively weak; the
corresponding IR band is strong. The medium-intense IR bands at
2731 and 2812 cm^{-1} are assigned to the Fermi resonance doublet of
ν_{C-H} and the first overtone of δ_{C-H} (of H-C=O group). The corres-

ponding Raman bands have weak band intensities. The strong IR band at 828 cm^{-1} assigned as γ_{C-H} (of H-C=O group) is relatively weak in the Raman spectrum.

The correlation rules for $\nu_{C=O}$ frequencies apply equally well for both IR and Raman, providing, of course, that the particular C=O vibration in question is IR- and/or Raman-active. (See Secs. III.K and L.) However, as demonstrated above, the IR consistently gives a strong band in contrast to a weak or medium-intense Raman band.

C. C-S, S-H, and S-S Groups

There is a distinct advantage of Raman spectroscopy over IR spectroscopy in the field of organosulfur chemistry where direct observation of C-S, S-S, or S-H is concerned. In the IR, the C-S and S-S vibrations in alkyl and/or arylsulfides and disulfides have weak band intensities and are frequently masked by strong out-of-plane ring vibrations. Also, the S-H stretching absorption band has relatively weak intensity. In contrast, the C-S, S-S, and S-H Raman bands have high intensity. The IR and Raman spectra of allyl mercaptan are compared in Fig. 25. The S-H stretching absorption band (ν_{S-H}) at 2580 cm^{-1} is readily apparent in the Raman spectrum but relatively weak in the Ir. Figure 26, a summary of Data for ν_{S-H} in several organomercaptans, shows that ν_{S-H} occurs in the region 2585 to 2650 cm^{-1}. The ν_{S-H} frequencies are sensitive to change in phase. For example, ν_{S-H} in n-butylmercaptan occurs at 2580 cm^{-1} in solution and at 2560 cm^{-1} in the pure liquid phase. The Raman band for ν_{S-H} is intense and polarized. Sheppard and others [29] have correlated carbon-sulfur stretching (ν_{C-S}) with branching on the sulfur α-carbon atom of mercaptans. Figure 27 shows that ν_{C-S} for CH_3-S occurs in the region 705 to 785; RCH_2-S, 660 to 670; R_2CH-S, 600 to 630; and R_3C-S, 600 to 570 cm^{-1}. All of the Raman bands are intense and polarized. In contrast, the corresponding IR bands are weak. The ν_{C-S} vibration often appears as a doublet due to the existence of rotational isomers. For example, in the case of

1,2-dimercaptopropane, the bands at 613 and 622 cm^{-1} are assigned to ν_{C-S} rotational isomers of the secondary S-H group, while the bands at 729, 684, and 665 cm^{-1} are assigned to ν_{C-S} rotational isomers of the primary S-H group. Dimercaptans of form H-S-R-S-H, where the two CH_2-SH groups are separated by more than one carbon atom, exhibit only one ν_{C-S} band.

The IR and Raman spectrum of dimethylsulfide are compared in Fig. 28. In the Raman spectrum, the band at 745 cm^{-1} is depolarized and results from antisymmetric C-S-C stretching vibration, while the Raman band at 696 cm^{-1} is polarized and results from symmetric C-S-C stretching. In contrast to the Raman data, note how relatively weak the intrinsic intensities are for the corresponding IR bands.

Other organic sulfides exhibit both symmetric and antisymmetric stretching in the region 570 to 800 cm^{-1}. Figure 29 summarizes the data for the C-S-C stretching vibrations for several organic sulfides. The antisymmetric C-S-C vibration is somewhat difficult to assign because it occurs in the same region as does the CH_2 rocking vibration for straight-chain aliphatic hydrocarbon groups. In addition, the spectra are complex in the region 600 to 800 cm^{-1} due to the fact that these molecules exist as rotational isomers. For example, diethyl sulfide in the liquid phase shows IR bands assigned as $\nu_{antisymmetric\ C-S-C}$ rotational isomers at 762, 781, and 694 cm^{-1}, and IR bands at 652, 638, and 694 cm^{-1} assigned as $\nu_{symmetric\ C-S-C}$ rotational isomers. In the solid phase IR spectrum of diethylsulfide, only one isomer is observed which, according to Scott, has C_{2v} symmetry [30]. Branching at the α-carbon atom affects the $\nu_{symmetric\ C-S-C}$ vibrations. For example, $\nu_{symmetric\ C-S-C}$ in dimethylsulfide and di-tert-butylsulfide occur at 693 and 587 cm^{-1}, respectively, while $\nu_{antisymmetric\ C-S-C}$ for the same two compounds occur at 746 and 609 cm^{-1}, respectively.

The IR and Raman spectra of diphenylsulfide are compared in Fig. 30. The strong IR band at 690 cm^{-1} results from the well-known out-of-plane phenyl ring deformation, and completely obscures the $\nu_{symmetric\ \phi-S-\phi}$ absorption band. The intense 694-cm^{-1} Raman band results from the $\nu_{symmetric\ \phi-S-\phi}$ vibration.

IR and Raman spectra of organodisulfides are less complex in the 700 to 800 cm^{-1} region than the corresponding organosulfides, and more closely resemble the corresponding organomercaptans. These observations suggest that there is little or no coupling between the in- and out-of-phase C-S-S-C stretching vibration, and that both occur at the same frequency. Figure 31 is a summary of the data for S-S stretching vibrations and shows that this polarized band occurs in the region 500 to 550 cm^{-1}. The bands at 512 and 529 cm^{-1} in methylphenyl disulfide result from ν_{S-S} rotational isomers as demonstrated from variable temperature experiments.

D. Carbon-Carbon Double Bond Stretching, $\nu_{C=C}$

The carbon-carbon double bond stretching vibration is a well-known IR group frequency, but its band intensity varies from nil to strong depending upon the symmetry of the molecule and the substitution of nonpolar or polar groups on the C=C bond. For example, tetrachloroethylene has a center of symmetry, and by mutual exclusion shows no coincidence between the IR and Raman spectrum as shown in Fig. 32. In the IR, $\nu_{C=C}$ is inactive, while the strong Raman band at 1578 cm^{-1} results from $\nu_{C=C}$. The IR and Raman spectra of trans-1,2-dichloroethylene are compared in Fig. 33. Trans-1,2-dichloroethylene also has a center of symmetry. In this case, we do note a band centered at 1659 cm^{-1} in the vapor phase, but this band does not result from $\nu_{C=C}$ because $\nu_{C=C}$ is IR-inactive in trans-1,2-dichloroethylene. Moreover, we would expect $\nu_{C=C}$ for this compound to occur at a frequency nearer that for perchloroethylene (Fig. 32). The Raman band at 1582 cm^{-1} results from $\nu_{C=C}$ and the IR band at 1659 cm^{-1} results from the first overtone of the 838-cm^{-1} fundamental of trans-1,2-dichloroethylene. The IR and Raman spectra of vinyl chloride are compared in Fig. 34. Vinyl chloride has only a plane of symmetry. Therefore, its 12 fundamental vibrations are both IR- and Raman-active. In this case, $\nu_{C=C}$ at $\sim$1610 cm^{-1} has high band intensity in both spectra. The IR and Raman spectra of 1-octene are compared in Fig. 35. In this case, the medium-intense

IR band at 1642 cm^{-1} results from $\nu_{C=C}$, as does the strong Raman band at 1647 cm^{-1}. The IR and Raman spectra of methylacrylate are compared in Fig. 36. The IR bands at 1635 and 1620 cm^{-1} have been assigned as $\nu_{C=C}$ and the first overtone of the 810-cm^{-1} fundamental, respectively [4]. However, we suggest that these two bands both result from $\nu_{C=C}$ of rotational isomers, because the Raman bands at 1638 and 1628 cm^{-1} show a significant change in their band intensity ratio at liquid nitrogen temperatures. The 1638-cm^{-1} band is the predominant isomer at room temperature. Structures such as (general

for acrylates) [structures shown] and [structure shown] are two of the

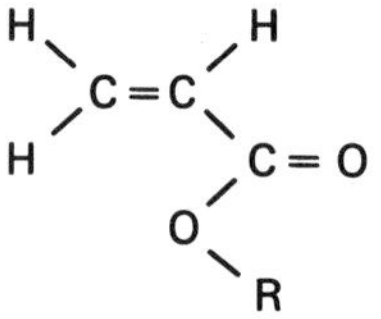

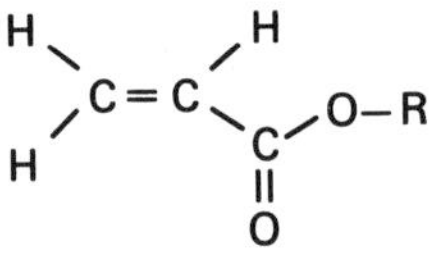

possible isomers. The IR and Raman spectra of ethyl fumarate are compared in Fig. 37. The IR band at 1648 cm^{-1} has been assigned as $\nu_{C=C}$. However, we note in the Raman spectrum that bands occur at 1650 and 1677 cm^{-1} with the higher-frequency band having more intensity than the lower-frequency band. In the solid phase, fumarates exhibit only one Raman band in this region, and it occurs at 1661 cm^{-1} in the Raman spectrum of dimethyl fumarate. These data indicate that dialkyl fumarates exist as rotational isomers and that in the IR only the $\nu_{C=C}$ for the isomer which doesn't have a center of symmetry is observed, while in the Raman, rotational isomers with and without a center of symmetry are observed:

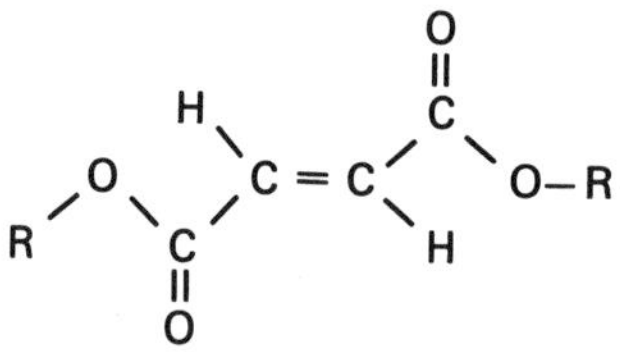

(A) center of symmetry   (B) center of symmetry

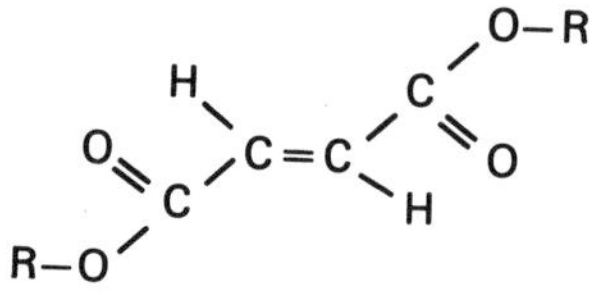

and [structure shown]

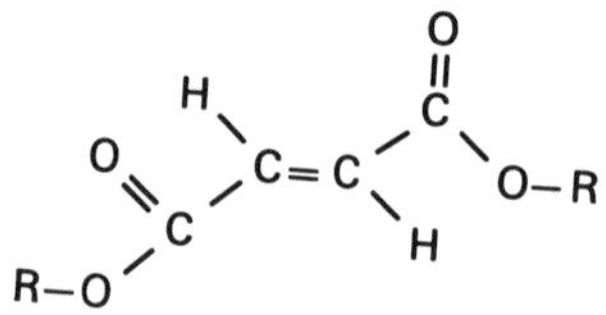

(C) no center of symmetry

Therefore, in the Raman, $\nu_{C=C}$ for isomers such as (A), (B), and (C) would be observed, but in the IR only $\nu_{C=C}$ for a rotational isomer such as (C) would be observed.

E. Triple Bond Stretching, $\nu_{C\equiv C}$ and $\nu_{C\equiv N}$

The IR and Raman spectra of diphenyl butadiyne are compared in Fig. 38. The IR bands near 2200 cm^{-1} are weak, and it is not apparent that this molecule contains C≡C groups. However, the intense Raman band at 2221 cm^{-1}, $\nu_{C\equiv C}$, leaves no doubt that the molecule contains C≡C. The $\nu_{C\equiv C}$ vibration has low IR band intensity unless substituted directly with a polar group or atom, but the Raman band is always intense.

The nitrile group is usually, but not always, readily detected in the IR spectrum. The IR and Raman spectra of 2,4,4,4-tetrachloro-butyronitrile are compared in Fig. 39. In the IR, the C≡N stretching vibration is always extremely weak for compounds containing halogen atoms in the α- or 2-position to the nitrile group. Even though this sample of 2,4,4,4-tetrachlorobutyronitrile fluoresced, the intense Raman band at 2250 cm^{-1} leaves no doubt that the molecule contains a nitrile group. The IR and Raman spectra of fumaronitrile are compared in Fig. 40 and the IR and Raman spectra of maleonitrile are compared in Fig. 41. The C≡N group can be readily identified from either IR or Raman for both compounds. However, it should be noted that the C=C group is not apparent in the IR spectrum of fumaronitrile, because this molecule has a center of symmetry and the $\nu_{C=C}$ vibration is IR-inactive. The $\nu_{C=C}$ stretching vibration is readily apparent in the Raman spectra near 1600 cm^{-1} for both of these molecules. It should be noted here that the well-known out-of-plane hydrogen deformations which give intense IR bands at 968 cm^{-1} for the trans structure in fumaronitrile and at 769 cm^{-1} for the cis structure in maleonitrile are either Raman-inactive, as is the case for fumaronitrile, or extremely weak, as is the case for maleonitrile.

F. Organophosphorus Compounds

Organophosphorus compounds are often used as insecticides and
are an important class of agricultural chemicals. In a recent book
edited by Halmann, the vibrational spectroscopy of phosphorus compounds is summarized by Nyquist and Potts in a systematic manner [28].
The IR and Raman spectral data are most useful in the elucidation of
the chemical structures of phosphorus compounds. The structures of
organophosphorus compounds tend to be complex, and the interpretation
of their spectra is usually difficult. However, the interpretation
can be greatly simplified if the problem and data are handled in a
systematic manner. In Fig. 42 a summary of the IR group frequency
correlations for organophosphorus compounds is given. These data
are useful in the determination of the chemical structures of phosphorus compounds. However, the combination of both IR and Raman
data is much more powerful in structure identification than just the
sum of each set of data. Spectroscopic data on model molecules are
often necessary in order to interpret the vibrational spectra of
more complex molecules. Thiophosphoryl dichloride fluoride has been
used as such a model compound for interpreting the vibrational spectra of compounds of form $R-O-P(=S)Cl_2$, whose spectra are much more
complex than would be expected [10,28,31,32]. The IR and Raman
spectra of $P(=S)Cl_2F$ are compared in Fig. 43. There is no difficulty
in assigning the P-F stretching vibration to the band at 900 cm^{-1},
because this mode would be expected to have strong IR band intensity
and weak Raman band intensity. The weak Raman band is expected because it is more difficult to deform the electron cloud about the
P-F bond than the P=S or PCl_2 bonds. The band near 740 cm^{-1} results
from P=S stretching, while the bands near 570 and 477 cm^{-1} are due
to the asymmetric and the symmetric PCl_2 stretching vibrations, respectively. Note how the asymmetric PCl_2 stretching vibration has
more band intensity than the symmetric PCl_2 stretching vibration in
the IR while in the Raman the band intensities are just reversed.
The IR and Raman spectra of O-methyl phosphorodichloridothioate are
compared in Fig. 44. In particular, the bands near 470, 545, and

715 cm^{-1} are doublets, and result from $\nu_{P=S}$, $\nu_{asym. \ PCl_2}$, and $\nu_{sym. \ PCl_2}$, respectively. Variable temperature studies have shown that the band intensity ratio of each doublet changes with change in temperature, and for this reason these bands are assigned to rotational isomers, resulting from the rotation of the C-O bond about the P-O bond [32].

It is known that organophosphorus compounds containing the group R-O-P(=S) intermolecularly rearrange to form R-S-P(=O) when heated to elevated temperatures. The IR and Raman spectra of S-methyl phosphorodichloridothioate are compared in Fig. 45. S-methyl phosphorodichloridothioate was prepared by heating O-methyl phosphorothioate at elevated temperatures. Again, bands near 1270, 700, 590, 550, and 460 cm^{-1} are doublets due to the existence of rotational isomers, the result of rotation of the C-S bond about the P-S bond. The $\nu_{P=O}$ doublet near 1270 cm^{-1} has high IR band intensity, but low Raman band intensity. The most intense Raman doublet occurs near 460 cm^{-1} and results from symmetric PSCl$_2$ stretching, while the doublet near 590 and 550 cm^{-1} results from a' and a'' antisymmetric PSCl$_2$ stretching. The doublet near 700 cm^{-1} results from C-S stretching [33].

The P=S group exhibits variable IR band intensity. In compounds such as 0,0-dimethyl S-[4-oxo-1,2,3-benzotriazin-3(4H)-yl methyl]-phosphorodithioate, both the IR and Raman band intensities are high for $\nu_{P=S}$. Note $\nu_{P=S}$ at 650 cm^{-1} in Fig. 46. The IR and Raman spectrum for 0,0-diethyl O-(3,5,6-trichloro-2-pyridyl)phosphorothioate are compared in Fig. 47. The strong Raman band at 640 cm^{-1} results from $\nu_{P=S}$. However, the corresponding IR band is noticeably weak [28].

G. NO$_2$ and SO$_2$ Groups

Although NO$_2$ groups substituted on carbon are usually readily detected from the IR spectrum, there are cases where it would be helpful to have additional information in order to verify the presence of these groups. The IR and Raman spectra of nitrobenzene are

compared in Fig. 48. The strong IR bands at $\sim$1530 and $\sim$1353 cm^{-1}
result from antisymmetric and symmetric -NO$_2$ stretching, respectively.
The most intense Raman band at 1352 cm^{-1} results from symmetric NO$_2$
stretching, while the Raman band at 1530 cm^{-1} assigned to the antisymmetric NO$_2$ stretching vibration is very weak. The IR and Raman
spectra of O,O-dimethyl O-(3-methyl-4-nitrophenyl)phosphorothioate
are compared in Fig. 49. The IR and Raman bands for antisymmetric
and symmetric NO$_2$ stretching vibrations in this compound are comparable to those for nitrobenzene. The Raman band intensity for symmetric NO$_2$ stretching is very intense, in fact, more intense than
$\nu_{P=S}$ near 650 cm^{-1}. The combination of IR and Raman data make it
virtually impossible to miss identification of the aryl-NO$_2$ group in
organic molecules.

A group such as -SO$_2$- is also usually readily detected from the
IR spectrum. The IR and Raman spectra of diphenylsulfone are compared in Fig. 50. The strong IR bands at 1324 and 1161 cm^{-1} result
from antisymmetric and symmetric SO$_2$ stretching, respectively. The
weak Raman band at 1315 cm^{-1} and the medium-intense Raman band at
1159 cm^{-1} result from antisymmetric and symmetric SO$_2$ stretching,
respectively.

H. Carbon-Halogen Stretching

The IR and Raman spectra for 1-chloro-4-fluorobutyne-2 and
1-bromo-4-chlorobutyne-2 are compared in Figs. 51 and 52, respectively. The intense Raman band at 2250 cm^{-1} is due to the C≡C group.
The IR bands at 999 and 706 cm^{-1} in 1-chloro-4-fluorobutyne are
assigned as carbon-fluorine and carbon-chlorine stretching, respectively. It should be noted that the IR band intensity for ν_{C-F} is
higher than that for ν_{C-Cl} and ν_{C-Br}, and the IR band intensity for
ν_{C-Cl} is higher than that for ν_{C-Br}. Comparison of the corresponding
Raman bands for ν_{C-F}, ν_{C-Cl}, and ν_{C-Br} in these two compounds shows
that their band intensities increase progressively from C-F to C-Br;
this is just the reverse trend for the corresponding IR band intensities. Such band intensity behavior is just what is expected,

because the dipole moment for C-X stretching decreases progressively in the series C-F thru C-I, as it becomes easier to distort the electron cloud about the C-X bond in the series. It is well known that ν_{C-X} also decreases in frequency with increased branching on the α-carbon atom. The IR and Raman spectra for 2-bromopropane and 2-iodopropane are compared in Figs. 53 and 54, respectively. The intense band near 540 cm^{-1} results from ν_{C-Br} in 2-bromopropane, and the intense band near 490 cm^{-1} results from ν_{C-I} in 2-iodopropane.

I. Substituted Benzenes

The out-of-plane hydrogen deformations used to determine chemical substitution on aromatic rings behave in a predictable manner when the ring substituent is an ortho or para director. However, the deformations occur at a higher frequency with a variable intensity often weaker than normal when the substituent is a meta director, and this can cause difficulty in the identification of a compound. To illustrate this point, the IR and Raman spectra of chlorobenzene are compared in Fig. 55. The strong IR band at 740 cm^{-1} and the medium-intense IR band at 684 cm^{-1} result from the in-phase out-of-plane hydrogen deformation and the out-of-plane phenyl ring-puckering deformation, respectively. These bands are very characteristic for monosubstituted benzenes substituted with ortho and para directors (see Figs. 30, 46, 54). The IR and Raman spectra for nitrobenzene are compared in Fig. 48, and the IR spectrum in the region 650 to 1000 cm^{-1} does not readily indicate that the sample is a monosubstituted benzene. However, there is a set of Raman bands for monosubstituted benzenes that is essentially unaffected by whether the atom or group is meta-, para-, or ortho-directing. The Raman bands at 1023 (polarized, medium), 1002 (polarized, strong), and 613 (depolarized, weak) cm^{-1} in the spectrum of chlorobenzene are quite indicative of a monosubstituted benzene (compare Figs. 30, 46, 55 versus 20, 23, 24, 44). The normal coordinates for these vibrations in chlorobenzene are shown in Table 1 along with the calculated frequencies as determined by Scherer [34-36].

TABLE 1

Monosubstitution

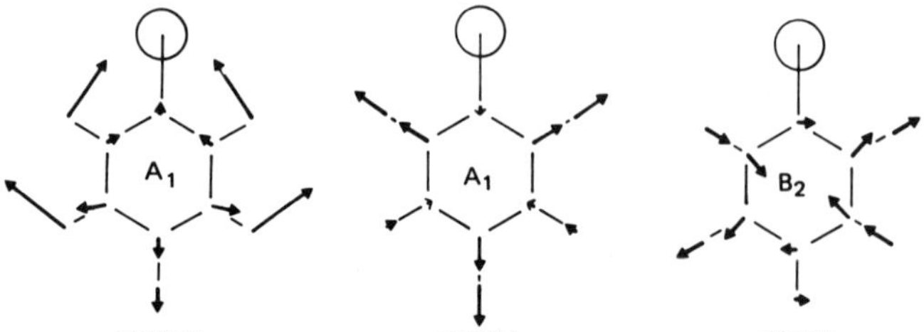

1023.84   995.54   620.35

Compound	Mode (cm^{-1})		
	A_1	A_1	B_2
ϕ-Cl	1023	1002	613
ϕ-Br	1022	1002	612
ϕ-I	1012	996	605
ϕ-OH	1029	1003	621
ϕ-NO$_2$	1027	1009	611
ϕ-CN	1024	1001	623
ϕ-OEt	1024	994	609
ϕ-Et	1027	1000	612
ϕCH$_2$CN	1024	999	609
ϕCH$_2$Cl	1024	998	611
ϕCOCH$_3$	1031	1007	623
ϕCOCl	1029	1008	611
ϕCOCH(CH$_3$)$_2$	1020	999	610
ϕ-CO-ϕ	1032	1005	622
ϕOCH$_2$CH=CH$_2$	1030	997	611
ϕCHNϕ	1022	1000	619
ϕ-O-ϕ	1019	999	609
ϕCH$_2$CHOHϕ	1027	1002	619
ϕ_2CH$_2$	1028	1000	620
ϕ_3CH	1028	998	615
ϕ_3SiH	1028	1000	621
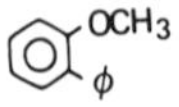 ⬡$^{OCH_3}_{\phi}$	1032	995	610

TABLE 1 (continued)

Compound	Mode (cm^{-1})		
	A_1	A_1	B_2
ϕ-SCH$_3$	1029	1003	619
ϕ-SO$_2$Cl	1024	1001	613
ϕ-SO-ϕ	1029	1001	618
ϕ-CH-CHϕ	1031	1003	623
Limits	1032-10	1010-994	625-603
Polarization	Polarized	Polarized	Dipolarized
Intensity	Medium	Strong	Weak

Quite clearly, none of these motions involve contributions from the substituent group. Table 1 summarizes the Raman group frequency data for 26 monosubstituted benzenes. The IR and Raman spectra of p-dichlorobenzene and p-nitrophenol are compared in Figs. 56 and 57, respectively. The strong band at 821 cm^{-1} in the IR spectrum of p-dichlorobenzene results from the in-phase out-of-plane hydrogen deformation. Other p-disubstituted benzenes with para- and ortho-directing atoms or groups show a similar characteristic absorption band. However, the IR spectrum of p-nitrophenol in the region 650 to 900 cm^{-1} is complex, and it isn't readily apparent from this spectrum that the compound is a p-disubstituted benzene derivative. Table 2 gives the frequencies for the characteristic Raman bands for seven p-disubstituted benzene derivatives. The normal coordinates [34-36] for these vibrations in p-dichlorobenzene are given in Table 2. It should be noted that the most intense Raman band in the spectrum of p-nitrophenol occurs at 1332 cm^{-1}, and results from the symmetric NO$_2$ stretching vibration. This is 20 cm^{-1} lower in frequency than $\nu_{sym. NO_2}$ occurs in nitrobenzene. The lower $\nu_{sym. NO_2}$ frequency in p-nitrophenol apparently results from intermolecular hydrogen bonding between OH and NO$_2$. The IR and Raman spectra for m-dichlorobenzene are compared in Fig. 58. The IR bands near 870, 785, and 674 cm^{-1} result from out-of-plane vibrations which are characteristic of m-disubstituted benzenes. The strong Raman band

TABLE 2

para Disubstitution

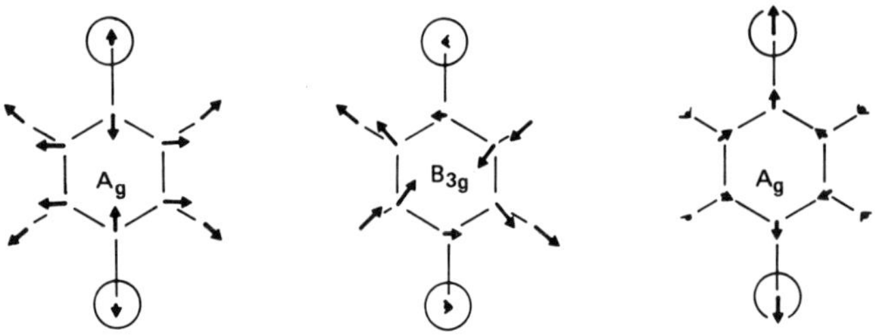

| | 746.18 | 626.74 | 328.02 |

Substituent	Mode (cm^{-1})		
	A_g	B_{3g}	A_g
Cl, OH	(823,835)	640	(336-389)
NO_2, CO_2H	(795-867)	630	332
CH_3, $CH(CH_3)_2$	800	640	300
Cl, Cl	748	630	330
CH_3, OCH_3	811	634	338
F, OCH_3	832	635	375
NO_2, OH	832	642	390
Intensity	Strong	Weak	Weak
Polarization	?	Dipolarized	?
Limits	?	630-642	?

at 1002 cm^{-1} in m-dichlorobenzene is characteristic of all meta-di-substituted benzenes, and Table 3 summarizes the data for seven meta-disubstituted benzenes. The normal coordinates for the characteristic band is also given in Table 3 [34-36].

The IR and Raman spectra for o-dichlorobenzene are compared in Fig. 59. The strong IR band at 749 cm^{-1} results from the in-phase out-of-plane hydrogen deformation, and is typical of o-disubstituted benzenes (with the exception of o-disubstituted benzenes containing

TABLE 3

meta Disubstitution

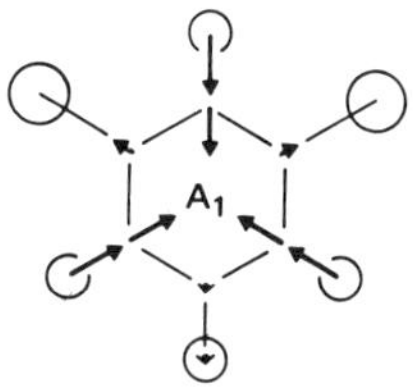

Substituent	Mode (cm^{-1}) 995.70
Cl, OH	1000
NO_2, NO_2	1008
CH_3, CH_3	998
F, F	1071
Cl, Cl	1002
CH_3, OCH_3	992
F, NO_2	1008
Intensity	Strong
Polarization	Polarized
Limits	992-1011

meta-directing groups). The IR and Raman spectrum of o-dinitroben-zene are compared in Fig. 60. The complexity of the IR spectrum in the region 650 to 900 cm^{-1} is not helpful in depicting that the sample is o-dinitrobenzene. There are two relatively strong Raman bands characteristic of ortho-disubstituted benzenes. These two Raman bands occur at 1041 and 662 cm^{-1} in o-dichlorobenzene. Table 4 gives the normal coordinates for these two in-plane vibrations [34-36] and also summarizes corresponding data for 12 other ortho-disubstituted benzene derivatives.

TABLE 4

ortho Disubstitution

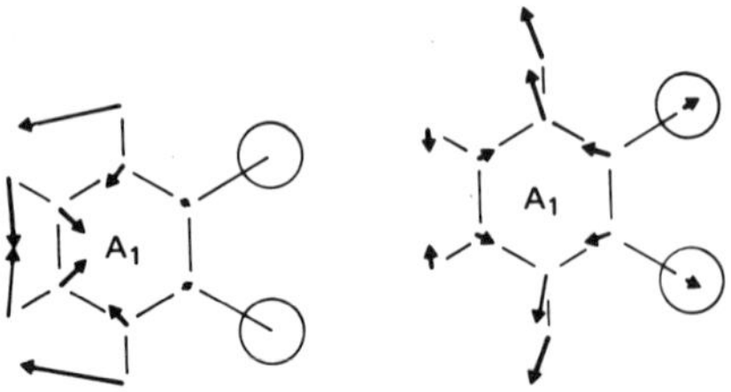

	1033.73	675.57
Substituent	Mode (cm^{-1})	
Cl, OH	1031	682
CH_3, CH_3	1051	734
Cl, CH_3	1047	679
Cl, Cl	1041	662
F, F	1026	762
CH_3, OCH_3	1053	745
Br, OCH_3	1030	655
I, OCH_3	1018	638
CO_2CH_3, CO_2CH_3	1039	650
OH, $COCH_3$	1037	715
NO_2, OH	1034	670
NO_2, NO_2	1040	643
NO_2, CO_2H	1045	647
Intensity	Strong	Strong
Polarization	Polarized	Polarized

J. Azines, C=N-N=C Stretching

Azines of the form (aryl-CH=N-)$_2$ can have the possible structures

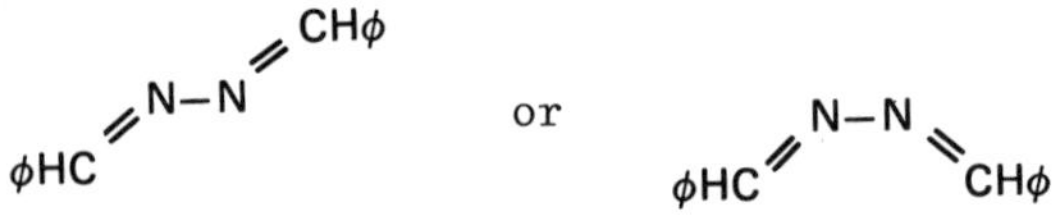

(A)   (B)

Structure (A) has a center of symmetry and would give rise to vibrations which are only IR-active and vibrations which are only Raman-
active. In particular, the out-of-phase stretching vibration

C=N–N=C   is IR-active only, while the in-phase stretching

vibration   C=N–N=C   is Raman-active only in (A). In (B),

both the out-of-phase   C=N–N=C   and in-phase   C=N–N=C

stretching vibrations are both IR- and Raman-active. In a study of
12 azines of form $(sub\text{-}\phi\text{-}CH=N\text{-})_2$, a strong IR band is observed in
the region 1598 to 1631 cm^{-1}, and a strong Raman band is observed
in the region 1538 to 1563 cm^{-1}. These bands are assigned to the
out-of-phase and in-phase $(C=N\text{-})_2$ stretching vibrations, respectively.
The IR and Raman spectra for azine:benzaldehyde are compared in Fig.
61. The strong IR band at 1625 cm^{-1} is assigned as the $(C=N\text{-})_2$ out-
of-phase stretching vibration, and the strong Raman band at 1556 cm^{-1}
is assigned as the $(C=N\text{-})_2$ in-phase stretching vibration. Since the
1556-cm^{-1} band is not apparent in the IR spectrum and the 1625-cm^{-1}
band is not apparent in the Raman spectrum, it is likely that azine:
benzaldehyde exists in a configuration such as (A) with a center of
symmetry rather than in a structure such as (B). The IR spectrum
clearly indicates that the sample contains monosubstitution. The
Raman spectrum in this case shows the 620-, 1000-, and 1025-cm^{-1}
bands, but the 1025-cm^{-1} band has less relative intensity than other
monosubstituted benzene derivatives.

K. Miscellaneous Raman Correlations

Raman correlations are not as well developed as IR correlations
[37]. Raman correlations for several classes of chemical compounds
are presented for reference in Figs. 62-76. The upper graph in each
figure represents the Raman group frequencies for that class of compounds. The band intensities are proportional to $45\alpha'^2 + 4\beta'^2$, where
α' and β' are the derivative of the mean polarizability and anisotropy
of the polarizability tensor, respectively, with respect to the normal
coordinate of the vibration. The horizontal lines across the range of
frequencies indicates the minimum band intensity for that particular
band in the series of compounds studied. The lower graph in each
figure (if given) represents the Raman group frequencies as measured,
and the intensities are proportional to $3\alpha'^2$. The number in paren-
theses after the name of the compound indicates the number of compounds used to develop the particular correlation.

Figure 62 shows a Raman correlation chart for alkyl benzoates.
Raman bands resulting from the alkyl group have not been included in
the correlation chart. The Raman bands at $\sim$620, 1010, and 1030 cm^{-1}
are characteristic of all monosubstituted benzenes, and were previously discussed (see Table 1). The band at $\sim$1722 cm^{-1} results from
C=O stretching, and the band at $\sim$3080 cm^{-1} results from phenyl hydrogen stretching.

Figure 63 shows a Raman correlation chart for dialkyl phthalates
[38]. Raman bands resulting from the alkyl group have not been included in the correlation chart. All ortho-disubstituted benzenes
exhibit a strong band in the region 1018 to 1053 cm^{-1}, and a medium
band in the region 638 to 762 cm^{-1}. The corresponding bands in
phthalates occur in the regions 1041 to 1047 cm^{-1} and 652 to 657 cm^{-1},
and both bands are polarized. Other bands in this correlation chart
arising primarily from vibrations within the orthophenylene group
have been compared with corresponding bands and assignments for
orthodichlorobenzene [38]. The C=O stretching vibration occurs in
the region 1728 to 1738 cm^{-1}.

Figure 64 shows a Raman correlation chart for alkylacetylricin-
oleate. Raman bands resulting from saturated aliphatic C-H stretching vibrations are not given. The Raman band at 3020 cm^{-1} results
from olefinic carbon-hydrogen stretching, the band at 1660 cm^{-1} results from carbon-carbon double bond stretching, and the band at
1740 cm^{-1} results from carbonyl stretching. The band at 1443 cm^{-1}
results from CH$_2$ bending, and for purpose of identification, it is
significant because it is the most intense Raman band in the $I_{/\!/}$
($48\alpha'^2 + 4\beta'^2$) spectra of alkyl acetyl ricinoleates, while the three
higher frequency bands just discussed are significantly weaker.
Note in the upper spectrum how $\nu_{C=C}$ and δ_{CH_2} are nearly of equal
intensity.

Figure 65 shows a Raman correlation chart for dialkyl 1,2,3,6-
tetrahydrophthalate (the bands resulting from saturated C-H stretching vibrations are not given). Raman bands 3080 (depolarized) and
3040 polarized cm^{-1} are assigned as out-of-phase and in-phase cis
olefinic carbon-hydrogen stretching, respectively. A carbonyl
stretching vibration is assigned at 1740 cm^{-1} and the carbon-carbon
double bond stretching vibration is assigned at 1660 cm^{-1}. The band
near 1455 cm^{-1} is assigned as CH$_2$ bending, and the band near 835 cm^{-1}
most likely results from a symmetric ring stretching vibration. The
1740-, 3080-, and 3040-cm^{-1} bands are proof that the structure is
not dialkyl 3,4,5,6-tetrahydrophthalate.

Figure 66 shows a Raman correlation chart for alkyl oleate.
(Bands assigned as aliphatic carbon-hydrogen vibrations are not
given.) The band near 3015 cm^{-1} results from olefinic carbon-hydrogen
stretching, the band near 1745 cm^{-1} from carbonyl stretching, and the
band near 1660 cm^{-1} from carbon-carbon double bond stretching. The
intense band near 1450 cm^{-1} results from CH$_2$ deformation.

Figure 67 shows a Raman correlation chart for alkyl acrylate.
The bands at 1620 and 1635 cm^{-1} are assigned to $\nu_{C=C}$ of rotational
isomers and have been discussed previously for methyl acrylate (see
Fig. 36). Bands near 3000, 3045, 3075, and 3020 cm^{-1} are assigned

to vibrations resulting from olefinic carbon-hydrogen stretching. Clearly the 3045-cm^{-1} band results from a symmetric C-H stretching vibration. The $\nu_{C=C}$ Raman band intensity is strong compared to the $\nu_{C=O}$ band intensity, and in the IR, the band intensity ratio is reversed (see Fig. 36).

Figure 68 shows a Raman correlation chart for vinyl esters (bands resulting from saturated carbon-hydrogen stretching vibrations are not given). The Raman bands near 3055, 3100, and 3130 cm^{-1} are assigned as olefinic carbon-hydrogen stretching vibrations. The 3055-cm^{-1} band clearly results from the symmetric C-H stretching vibration. The band near 1762 and the band near 1650 cm^{-1} are assigned as $\nu_{C=O}$ and $\nu_{C=C}$, respectively.

Figure 69 shows a Raman correlation chart for dialkyl itaconate (bands resulting from aliphatic C-H stretching vibrations are not given). The bands near 3120 and 1647 cm^{-1} are assigned as $\nu_{=CH_2}$ and $\nu_{C=C}$, respectively. The bands near 1735 and 1719 cm^{-1} are assigned as $\nu_{C=O}$ for the nonconjugated and conjugated ester groups, respectively.

Figure 70 shows a Raman correlation chart for trialkyl aconitate (bands resulting from aliphatic C-H stretching vibrations are not given). The bands near 3075 and 1660 cm^{-1} are assigned as $\nu_{=C-H}$ and $\nu_{C=C}$, respectively. The band near 1730 cm^{-1} is assigned as $\nu_{C=O}$. Bands for more than one type of C=O ester group are not observed in the Raman spectrum.

Figure 71 shows a Raman correlation chart for dialkyl fumarate. The $\nu_{C=C}$ vibrations have been previously discussed, and in the liquid phase exist as a doublet in the Raman spectrum giving bands near 1650 and 1677 cm^{-1}. The band near 3070 cm^{-1} results from olefinic carbon-hydrogen stretching, and the band near 1730 cm^{-1} results from a $\nu_{C=O}$ vibration.

Figure 72 shows a Raman correlation chart for dialkyl maleate. The bands near 3065 and 1650 cm^{-1} are assigned as olefinic carbon-hydrogen stretching and $\nu_{C=C}$, respectively, and the band near 1735 cm^{-1} results from $\nu_{C=O}$.

Figure 73 shows a Raman correlation chart for dialkyl adipate (bands for carbon-hydrogen stretching vibrations are not given). The band near 1740 cm^{-1} is assigned as $\nu_{C=O}$. The band near 1452 cm^{-1} is assigned as CH$_2$ bending, δ_{CH_2}.

Figure 74 shows a Raman correlation chart for acetyl trialkyl citrate. Bands assigned as carbon-hydrogen stretching vibrations are not given. The bands near 1740 and 1750 cm^{-1} are assigned to $\nu_{C=O}$ vibrations. The band near 1458 cm^{-1} are assigned as δ_{CH_2}.

Figure 75 shows a Raman correlation chart for dialkyl sebacate (bands assigned as carbon-hydrogen stretching vibrations are not given). Bands near 1740 and 1448 cm^{-1} are assigned as $\nu_{C=O}$ and δ_{CH_2}, respectively.

Figure 76 shows a Raman correlation chart for alkene-1 (bands assigned to aliphatic carbon-hydrogen stretching vibrations are not given). The band near 3285 cm^{-1} is assigned as $2\nu_{C=C}$ and the band near 1649 cm^{-1} is assigned as $\nu_{C=C}$. Bands near 3010 and 3092 cm^{-1} are assigned as olefinic carbon-hydrogen stretching vibrations. Very weak Raman bands near 918 and 995 cm^{-1} most likely result from vinyl wagging and twisting vibrations, respectively.

L. Miscellaneous Infrared Correlation Charts

Figure 77 shows an IR correlation chart for the out-of-plane hydrogen deformations for olefins and related compounds. Data used to prepare this chart are discussed in [39]. In compounds containing the CH=CH$_2$ group, a pair of strong bands are observed in the region 810 to 1020 cm^{-1}. The higher-frequency band is assigned to the twisting vibration and the lower-frequency band to the wagging vibration. Both bands are strong in the IR but weak in the Raman.

Figure 78 shows an IR correlation chart for compounds containing a terminal acetylenic group. Data used to prepare this chart are discussed in [40]. The acetylenic $\equiv$C-H group exhibits strong bands in the regions 3310 to 3320 and 610 to 655 cm^{-1}, and are assigned as C-H stretching and C-H bending vibrations, respectively.

The C≡C stretching vibration occurs in the region 2100 to 2150 cm^{-1}
with variable band intensity. However, in most cases $\nu_{C≡C}$ has weak
band intensity. Compounds containing the H-C≡C-C=O group exhibit a
strong $\nu_{C≡C}$ band intensity. Often a weak band will be observed in
the region 1210 to 1275 cm^{-1}, which is assigned as a first overtone
of the acetylenic C-H bending vibration. Weak bands in the region
900 to 960 cm^{-1} are assigned to ≡C-C stretching and bands near 300
and 200 cm^{-1} have been assigned to skeletal bending vibrations.

Figure 79 shows an IR correlation chart for intramolecularly
hydrogen-bonded phenols prepared from data given in [41]. The band
in the region 2800 to 3600 cm^{-1} results from O-H stretching, while
the band in the region 875 to 350 cm^{-1} results from the out-of-plane
O-H bending vibration, γ_{O-H}. The γ_{O-H} vibration is readily apparent
in the IR spectrum due to the fact that it is strong and has a unique-
ly wide half-band width. The correlation shows that as ν_{O-H} decreases
in frequency, γ_{O-H} increases in frequency. In phenols which are not
intramolecularly hydrogen-bonded, the vibration is OH torsion, and
the band occurs at lower frequency, and occurs at 300 cm^{-1} for phenol.

Figure 80 shows an IR correlation chart for thiol esters, thiol
carbonates, and related compounds. The chart was prepared from data
discussed in [42 and 43]. The correlations presented in the region
1625 to 1825 cm^{-1} are based on bands assigned as $\nu_{C=O}$. The weak pair
of bands near 2800 cm^{-1} for thiol formates (O=C-H group) is assigned
to C-H stretching and the first overtone of C-H bending in Fermi
resonance. The weak-medium band near 2580 cm^{-1} is assigned as ν_{S-H}
for the O=C-S-H group. Bands in the region 800 to 1300 cm^{-1} are

assigned to "X—$\overset{\overset{\displaystyle O}{\displaystyle \|}}{C}$—Y" stretching vibrations.

Figure 81 is an IR correlation chart for secondary amides and
carbamates. The bands near 3400 cm^{-1} result from N-H stretching and
the bands near 1700 cm^{-1} result from C=O stretching. The lower frequency range in each series (if given) results from the hydrogen-
bonded species [44,45].

Figure 82 shows an IR correlation chart for double and triple bonds. Bands occuring in the region 1900 to 2400 cm^{-1} are assigned to C≡C, C≡N stretching or asymmetric X=C=Y stretching. Fundamentals in this region have variable band intensities from weak to strong. The C=C stretching vibration occurs in the region 1570 to 1790 cm^{-1} with variable band intensity from weak to strong. Some C=O stretching vibrations are also included and their band intensities are always strong.

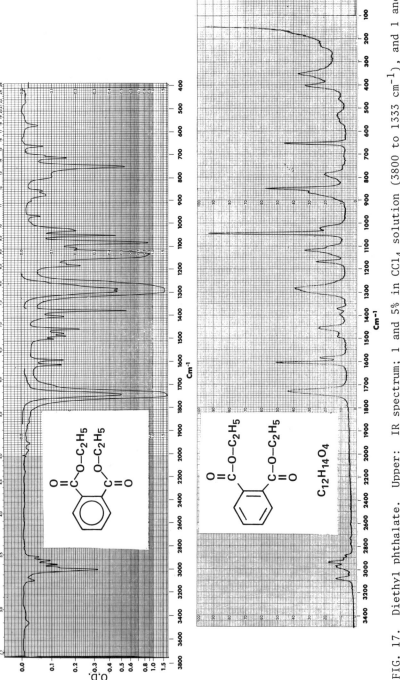

FIG. 17. Diethyl phthalate. Upper: IR spectrum; 1 and 5% in CCl₄ solution (3800 to 1333 cm⁻¹), and 1 and 5% in CS₂ solution (1333 to 450 cm⁻¹). Solvents are compensated. Lower: Raman spectrum; neat liquid.

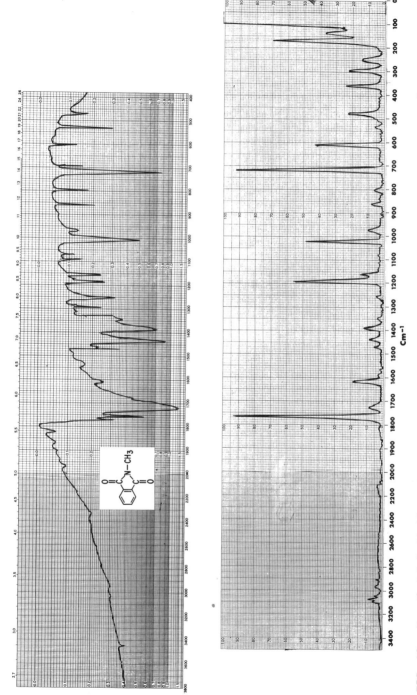

FIG. 18. N-methylphthalimide. Upper: IR spectrum; split mull [Fluorolube (3800 to 1333 cm^{-1}) and Nujol (1333 to 450 cm^{-1})]. Lower: Raman spectrum; neat solid.

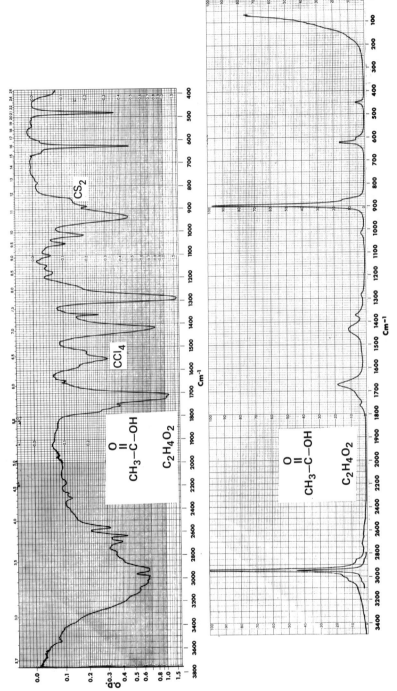

FIG. 19. Acetic acid. Upper: IR spectrum; 5% in CCl$_4$ solution (3800 to 1333 cm^{-1}), and 5% in CS$_2$ solution (1333 to 450 cm^{-1}). Lower: Raman spectrum; neat liquid. Solvents are not compensated.

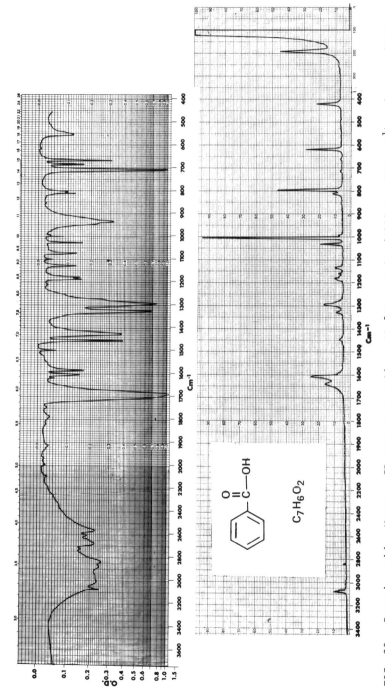

FIG. 20. Benzoic acid. Upper: IR spectrum; split mull [Fluorolube (3800 to 1333 cm^{-1}) and Nujol (1333 to 450 cm^{-1})]. Lower: Raman spectrum neat solid.

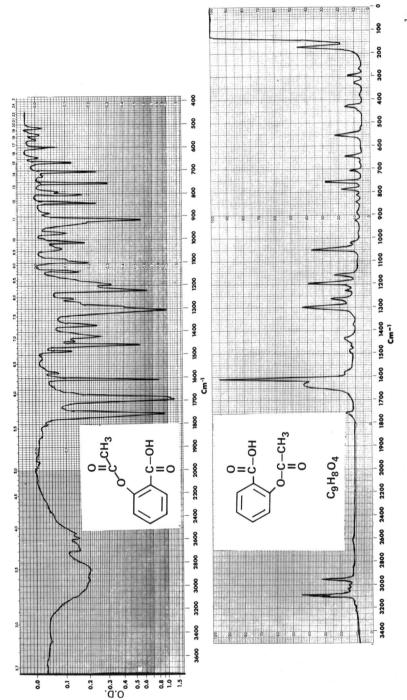

FIG. 21. Aspirin (acetyl salicylic acid). Upper: IR spectrum; split mull [Fluorolube (3800 to 1333 cm^{-1}) and Nujol (1333 to 450 cm^{-1})]. Lower: Raman spectrum; neat solid.

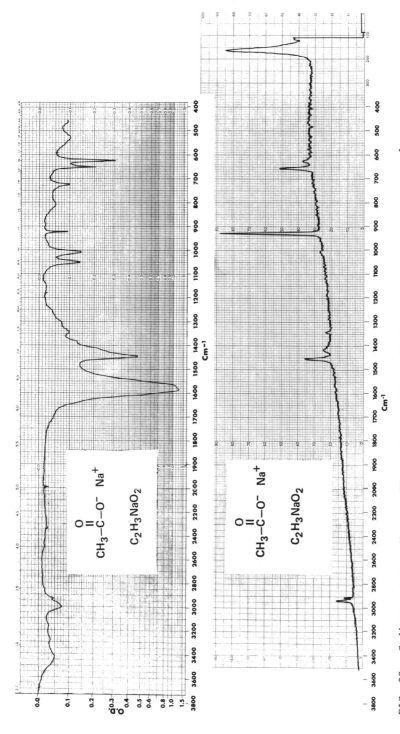

FIG. 22. Sodium acetate. Upper: IR spectrum; split null [Fluorolube (3800 to 1333 cm^{-1}) and Nujol (1333 to 450 cm^{-1})]. Lower: Raman spectrum neat solid.

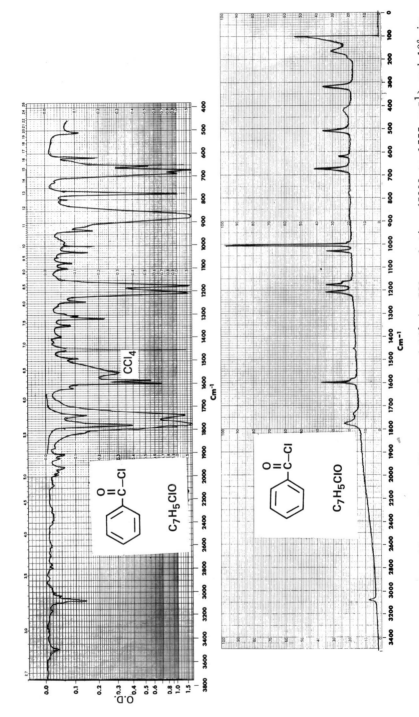

FIG. 23. Benzoyl chloride. Upper: IR spectrum; 1 and 10% in CCl₄ solution (3800 to 1333 cm⁻¹) and 10% in CS₂ solution (1333 to 450 cm⁻¹). Lower: Raman spectrum neat liquid.

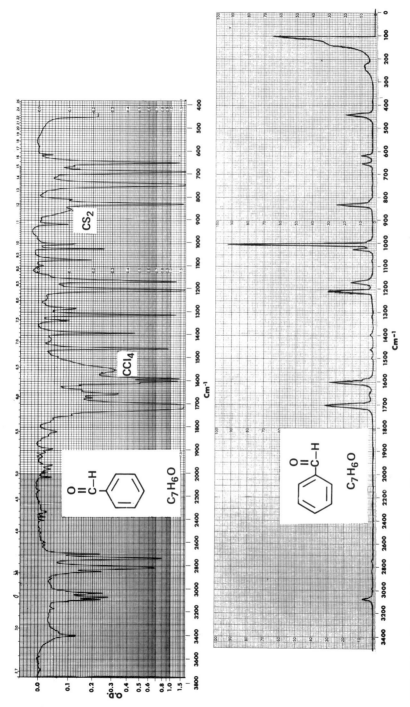

FIG. 24. Benzaldehyde. Upper: IR spectrum; 10% in CCl₄ solution and 10% in CS₂ solution. Lower: Raman spectrum; neat liquid.

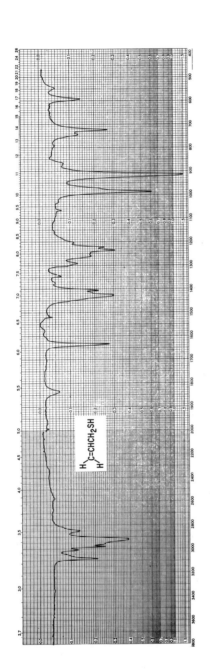

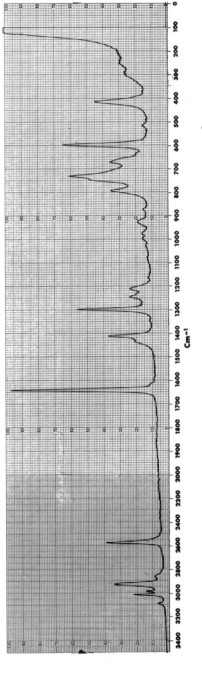

FIG. 25. Allyl mercaptan. Upper: IR spectrum; 10% in CCl$_4$ solution (3800 to 1333 cm^{-1}) and 10% in CS$_2$ solution (1333 to 4500 cm^{-1}). Solvents are compensated. Lower: Raman spectrum; neat liquid.

Sulfur-Hydrogen Stretching in Organomercaptans (cm^{-1})

R-SH	Infrared		Raman		
R	CCl_4 solution	liquid	liquid	R.I.	Dep. ratio
CH_3	2584	$\left\{\begin{matrix}2630\\2590\end{matrix}\text{ vapor}\right\}$			
iso-C_3H_7	2579	2562	2578	31.4	0.09
n-C_4H_9	2580	2560	2580	68.5	0.10
iso-C_4H_9	2583	2567	2582	61.3	0.09
sec-C_4H_9	2572	2559	2582	54.1	0.13
n-C_5H_{11}	2576	2565	2580	73.0	0.10
n-C_8H_{13}	2575	2560	2580	77.1	0.11
tert-$C_{12}H_{25}$	2579	2585	2583	100.0	0.13
$CH_2=CH-CH_2$	2574	2557	2575	72.5	0.10
$HOOC-CH_2$	-	2573	2583	100.0	0.13
$HOOC-CH_2-CH_2$	-	2575	2580	51.9	0.14
iso-$C_8H_{17}-OOC-CH_2$	2585	2577	2584	100.0	0.11
$C_2H_5-OOC-CH_2$	2583	2569	2579	93.3	0.12
HS-R-SH					
R					
C_2H_4	2569	2550	2566	100.0	0.24
$CH_3-CH-CH_2$	2565	2554	2579	100.0	0.09
n-C_3H_6	2580	2559	2571	90.5	0.12
n-C_4H_8	2580	2560	2571	75.5	0.12
n-C_5H_{10}	2580	2560	2582	92.7	0.12
n-C_6H_{10}	2581	2561	2571	87.9	0.11

FIG. 26. Summary of data for ν_{S-H} in organomercaptans.*
The relative intensities (R.I.) are based on ν_{S-H} or ν_{C-S}, whichever has the most intensity. The band intensity is actually stronger for the carbon-hydrogen stretching vibrations after C_6, and the intensity would be stronger yet for the carbon-hydrogen stretching vibrations with correction for the drop-off in energy of the laser.

Organomercaptans, R-S-H and HS-R-SH

R-SH	Infrared			Raman		
R	CS_2 soln cm^{-1}	neat cm^{-1}		neat cm^{-1}	R.I.*	Dep. ratio
CH_3	703	723 710 690	vapor			
$n-C_4H_9$	651	652		657	100.0	0.11
$n-C_5H_{11}$	653	654		657	100.0	0.14
$n-C_6H_{13}$	650	653		657	100.0	0.16
$iso-C_4H_9$	708 w / 668 vw	710 w / 668 vw		715 / 673	100.0 / 61.3	0.13 / 0.05
$iso-C_3H_7$	629 / 616 sh	631 / 620 sh		635 / 622 sh	100.0 / 61.0	0.09 / 0.10
$sec-C_4H_9$	608	610 / 619 sh		625 / 618 sh	100.0 / 85.0	0.13 / 0.14
$tert-C_4H_9$				587	100.0(2)	
$tert-C_{12}H_{25}$	632 w / 591 w	637 w / 588 w		630 / 599	83.9 / 99.5	0.13 / 0.16
HS-R-SH						
R						
CH_2-CH_2	635 / 667	635 / 667		641 / 675	89.5 / 20.0	0.15 / 0.68
$CH_3-CH-CH_2$	728 / 669 / 659 / 620 / 609	727 / 668 / 660 / 619 / 608		729 / 674 / 665 / 622 / 613	92.3 / 23.7 / 23.2 / 58.0 / 96.7	0.13 / 0.12 / 0.12 / 0.06 / 0.04

$CH_2-CH_2-CH_2$	665 651	664 654	669 sh 660 sh	100.0 83.9	0.08 0.13
$CH_2(CH_2)_2CH_2$	651	653	658	100.0	0.15
$CH_2(CH_2)_3CH_2$	653	653	661	100.0	0.15
$CH_2(CH_2)_4CH_2$	653	653	658	100.0	0.13

Other mercaptans R-S-H

R					
$CH_2-CH_2-CH_2$	725 743	725 734	731 745	100.0 74.4	0.29 0.24
$HOOCCH_2$	∿580 bd ∿680 bd		584 680	73.7 56.3	0.08 0.11
$C_2H_5-OOC-CH_2$	583 700	578 695	586 703	100.0(a) 28.1	0.07 0.36
$iso-C_8H_{17}-OOC-CH_2^-$	582 700	580 700	591 710	70.6 35.4	0.07 0.30
$(CH_2-OOC-CH_2)_2$	588 698		589 706	89.4 31.7	0.08 0.30
$HOOC-CH_2-CH_2$	∿650		678	100.0	0.16
$O(-CH_2-CH_2-)_2$	665	665	671	100.0	0.15

FIG. 27. Summary of C-S stretching frequencies for mercaptans. (See Fig. 26 legend for explanation of *.)

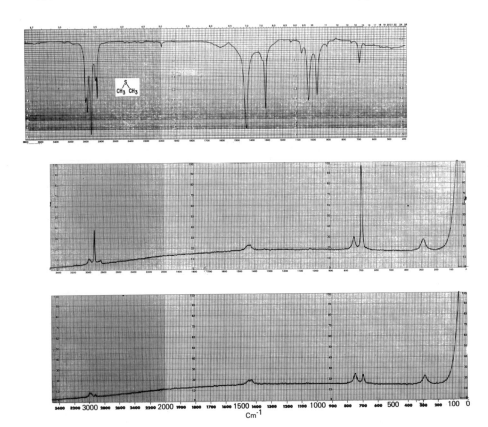

FIG. 28. Dimethyl sulfide. Upper: IR spectrum; 10% in CCl_4 solution (3800 to 1333 cm^{-1}) and 10% in CS_2 solution (1333 to 450 cm^{-1}). Solvents are compensated. Middle: Raman spectrum; neat liquid. Lower: polarized Raman spectrum; neat liquid.

Organosulfides, R-S-R'

R-S-R'		Infrared				Raman					
R	R'	νasym C-S-C CS$_2$ soln	νasym C-S-C neat	νsym C-S-C CS$_2$ soln	νsym C-S-C neat	νasym C-S-C neat	R.I.	Dep. ratio	νsym C-S-C	R.I.	Dep. ratio
CH$_3$	CH$_3$	743	746	691	693	745	15.3	0.81	696	100.0	0.11
CH$_3$	C$_2$H$_5$	729	729 720[b]		652 672 650[b]	732	43.8	0.75	661 684	100.0 40.8	0.19[a] 0.25
C$_2$H$_5$	C$_2$H$_5$	761 782 694	762 781 694 693[c]	651 637 694	652 638 694 693[c]	766 782 696	5.9 5.7 55.8	0.29 0.29 0.66	651 641 696	100.0 80.3 55.8	0.25 C$_1$ 0.23 C$_2$ 0.66 C$_{2v}$
C$_2$H$_5$	n-C$_4$H$_9$	743	743	651	661	750	28.4	0.72	660 643 694	100.0 38.6 41.5	0.16 0.19 0.54
n-C$_4$H$_9$	n-C$_4$H$_9$	745	746	650	660	750	78.6	0.47	659 670 688	100.0 81.3 48.6	0.25 0.29 0.5

FIG. 29. Summary of C-S-C stretching frequencies for organosulfides. [a]Low temperature isomer. [b]Band at 759 cm^{-1} presumably results from CH$_2$ rocking - Solid phase. [c]Band at 799 cm^{-1} presumably results from CH$_2$ rocking C$_{2v}$ - Solid phase.

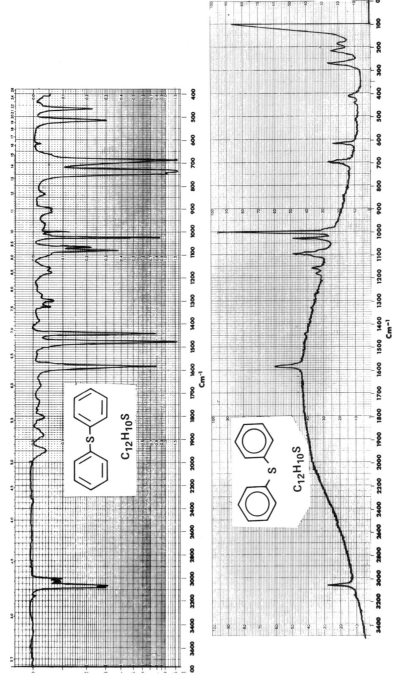

FIG. 30. Diphenyl sulfide. Upper: IR spectrum; 10% in CCl$_4$ solution (3800 to 1333 cm^{-1}) and 10% in CS$_2$ solution (1333 to 450 cm^{-1}). Solvents are compensated. Lower: Raman spectrum; neat liquid.

R-S-S-R'		Infrared	Raman		
R	R'	ν_{S-S}	ν_{S-S}	R.I.	Dep. ratio
CH_3	CH_3	511	511	100	0.06
C_2H_5	C_2H_5	511	511[d]	98.0[c]	0.06
		526	525	49.2[c]	0.12
iso-C_4H_9	iso-C_4H_9	512	516[d]	58.8[c]	0.06
		528	528	37.5[c]	0.14
tert-C_4H_9	tert-C_4H_9	502	515	7.2[c]	0.09
⬡-CH_3	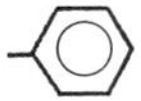 CH_3	507	512[e]	24.8[a]	0.22
		525	529[e]	23.0[a]	0.21
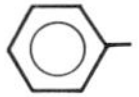 ⬡	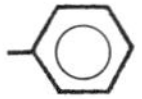 ⬡ (S=C)	b	546 (solid)	26.8[a]	–
C_2H_5-O-C(=S)-	-C(=S)-O-C_2H_5	501	501	100	0.08

FIG. 31. Summary of S-S stretching frequencies for organodisulfides (cm^{-1}). [a]Based on 1001 cm^{-1} ϕ ring mode as 100%. [b]Not observed in the IR spectrum. [c]Based on ν_{C-S} mode as 100%. [d]Predominate isomer at room temperature. [e]Predominate isomer at dry-ice temperature.

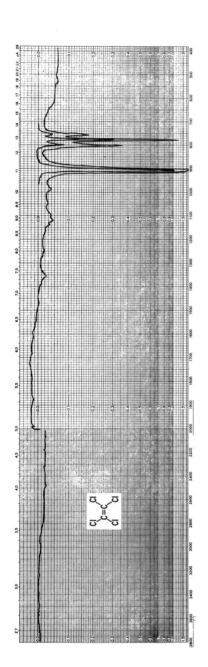

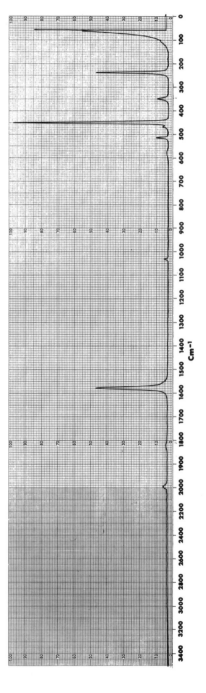

FIG. 32. Tetrachloroethylene (perchloroethylene). Upper: IR spectrum; 10% in CCl$_4$ solution (3800 to 1333 cm^{-1}) and 2 and 10% in CS$_2$ solution (1333 to 450 cm^{-1}). Solvents are compensated. Lower: Raman spectrum; neat liquid.

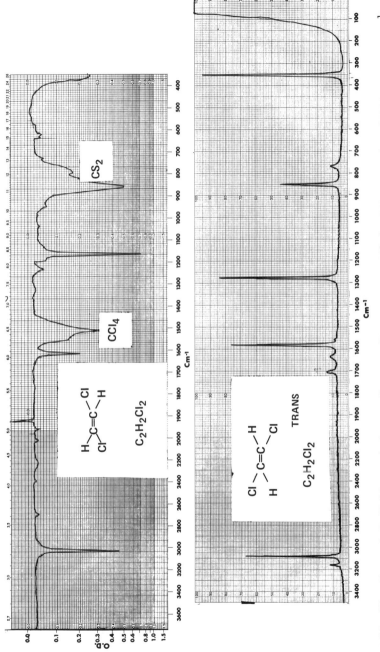

FIG. 33. Trans-1,2-dichlorethylene. Upper: IR spectrum; 10% in CCl₄ solution (3800 to 1333 cm⁻¹) and 10% in CS₂ solution (1333 to 450 cm⁻¹). Solvents are not compensated. Lower: Raman spectrum; neat liquid.

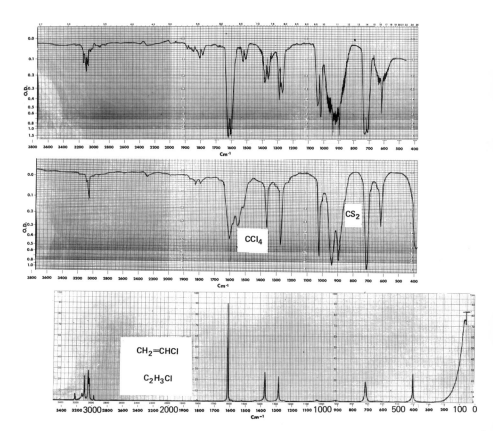

FIG. 34. Vinyl chloride. Upper: IR spectrum; vapor at 120-mm Hg in a 10-cm cell. Middle: IR spectrum; 10% in CCl$_4$ solution (3800 to 1333 cm^{-1}) and 10% in CS$_2$ solution (1333 to 450 cm^{-1}). Solvents are not compensated. Lower: Raman spectrum; neat liquid.

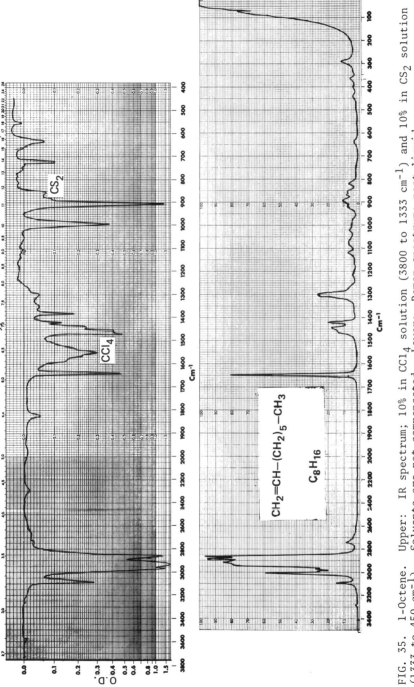

FIG. 35. 1-Octene. Upper: IR spectrum; 10% in CCl$_4$ solution (3800 to 1333 cm^{-1}) and 10% in CS$_2$ solution (1333 to 450 cm^{-1}). Solvents are not compensated. Lower: Raman spectrum; neat liquid.

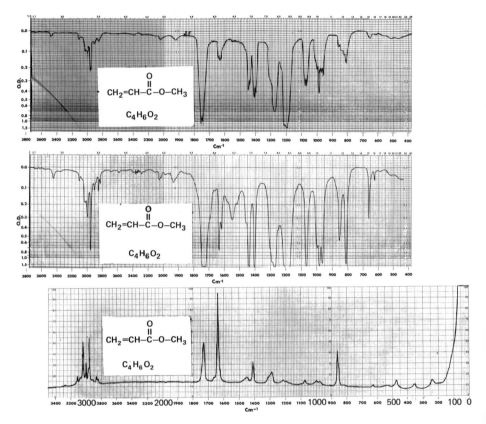

FIG. 36. Methyl acrylate. Upper: IR spectrum; vapor at 25-mm Hg
sample to 600-mm Hg with N_2 in a 5-cm cell. Middle: IR spectrum;
10% in CCl_4 solution (3800 to 1333 cm^{-1}) and 10% in CS_2 solution
(1333 to 450 cm^{-1}). Lower: Raman spectrum; neat liquid.

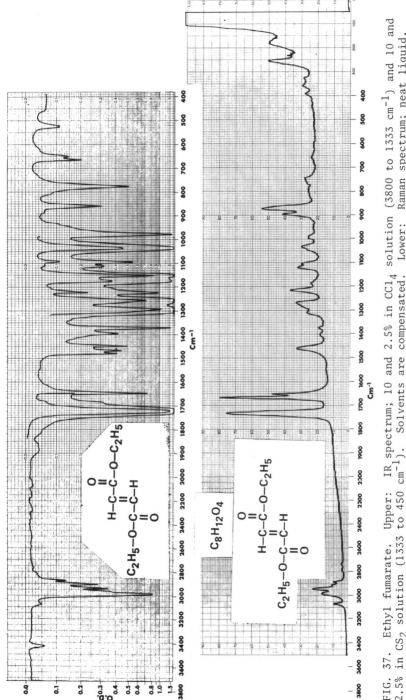

FIG. 37. Ethyl fumarate. Upper: IR spectrum; 10 and 2.5% in CCl₄ solution (3800 to 1333 cm⁻¹) and 10 and 2.5% in CS₂ solution (1333 to 450 cm⁻¹). Lower: Raman spectrum; neat liquid. Solvents are compensated.

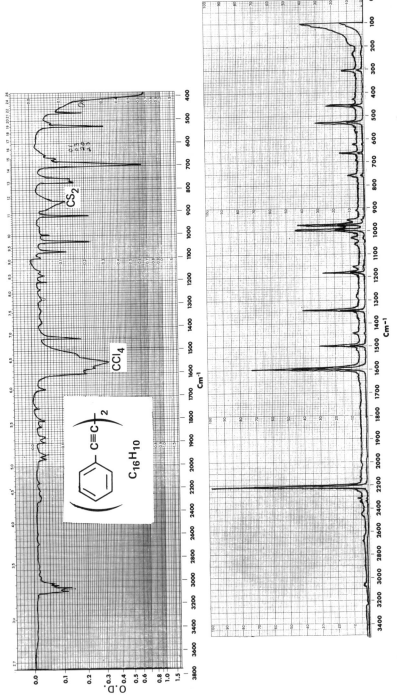

FIG. 38.   Diphenyl butadiene.   Upper: IR spectrum; 5% in CCl$_4$ solution (3800 to 1333 cm^{-1}) and 5% in CS$_2$ solution (1333 to 450 cm^{-1}).   Solvents are not compensated.   Lower: Raman spectrum; neat solid.

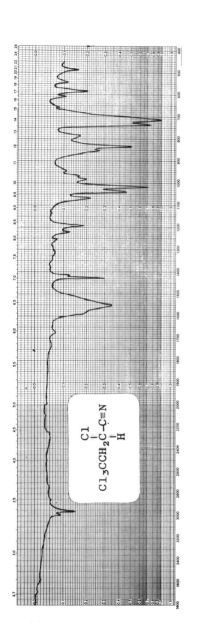

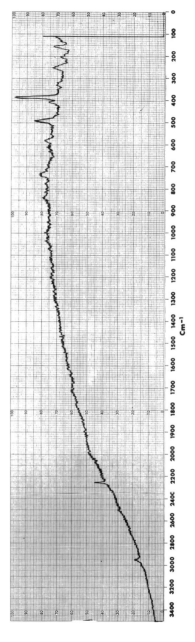

FIG. 39. 2,4,4,4-tetrachlorobutyronitrile. Upper: IR spectrum; 10% in CCl₄ solution (3800 to 1333 cm⁻¹) and 10% in CS₂ solution (1333 to 450 cm⁻¹). Solvents are not compensated. Lower: Raman spectrum; neat liquid. The sample is fluorescing, and is the cause of the high background.

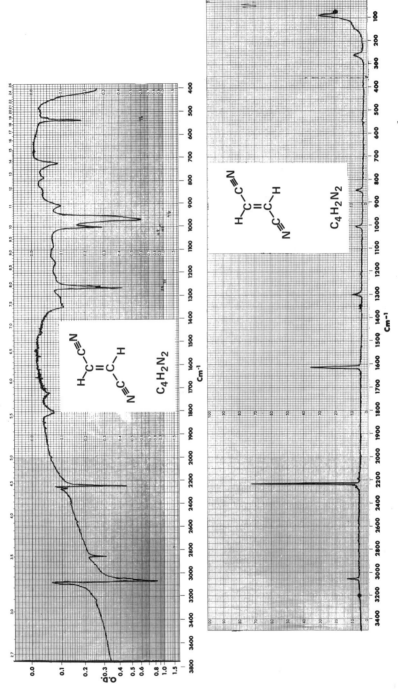

FIG. 40. Fumaronitrile. Upper: IR spectrum; split mull [Fluorolube (3800 to 1333 cm⁻¹) and Nujol (1333 to 450 cm⁻¹)]. The bands above 2200 cm⁻¹ are perturbed due to the Christiansen Effect. Lower: Raman spectrum; neat solid.

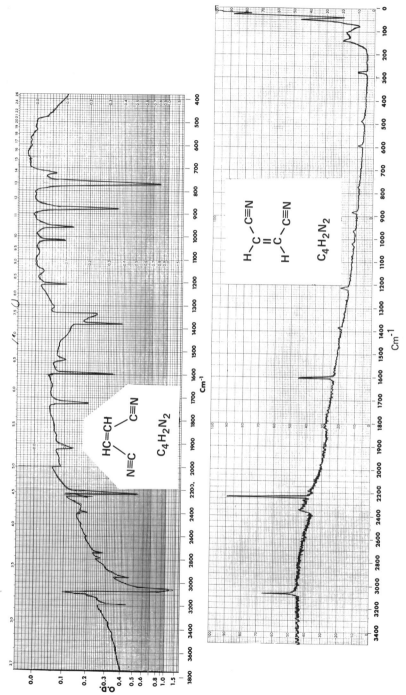

FIG. 41. Maleonitrile. Upper: IR spectrum; split mull [Fluorolube (3800 to 1333 cm^{-1}) and Nujol 1333 to 450 cm^{-1})]. The bands above 2200 cm^{-1} are perturbed due to the Christiansen Effect. Lower: Raman spectrum; neat solid.

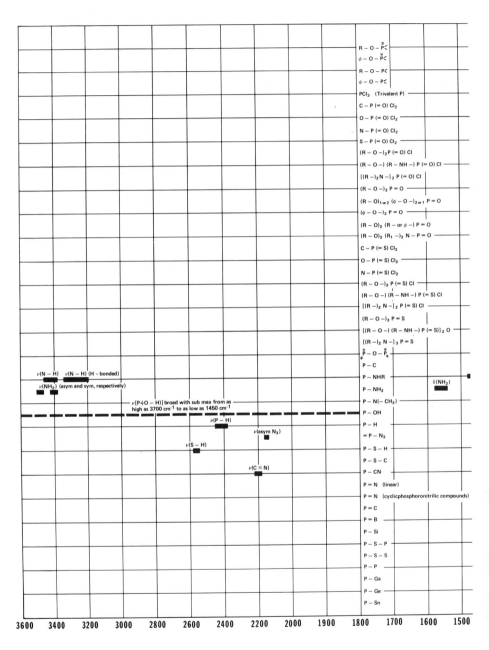

FIG. 42. IR group frequency correlations for organophosphorus compounds.

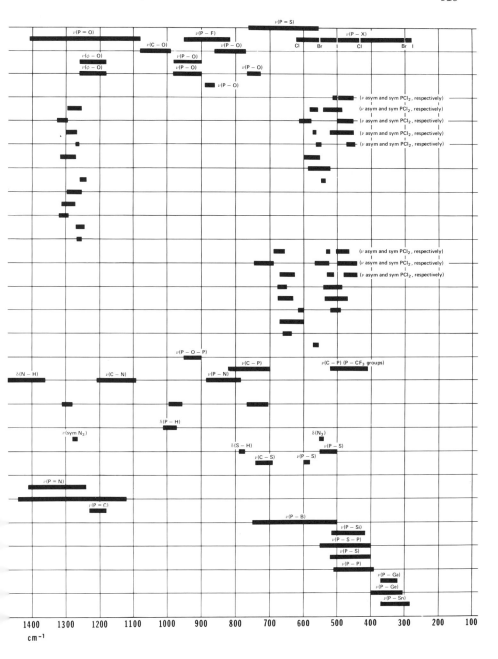

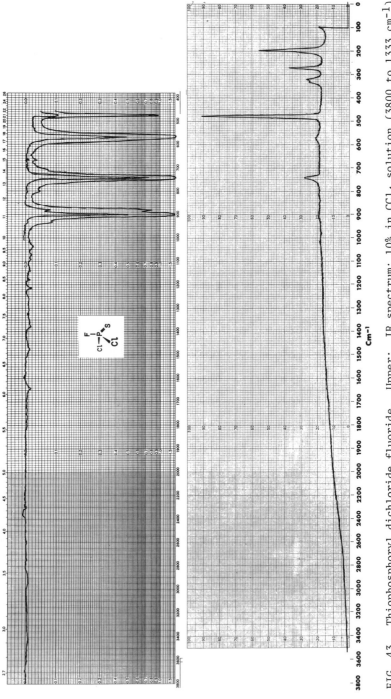

FIG. 43. Thiophosphoryl dichloride fluoride. Upper: IR spectrum; 10% in CCl$_4$ solution (3800 to 1333 cm^{-1}) and 10% in CS$_2$ solution (1333 to 450 cm^{-1}). Solvents are compensated. Lower: Raman spectrum; neat liquid.

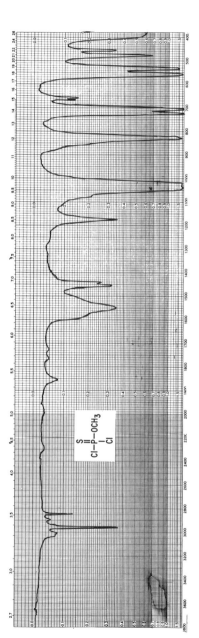

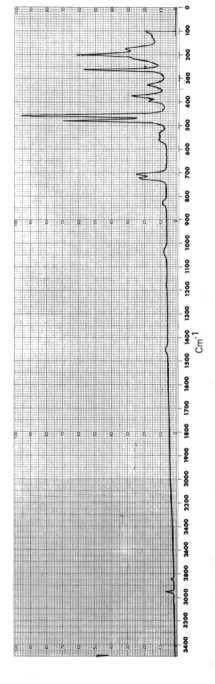

FIG. 44. O-Methyl phosphorodichlorothioate. Upper: IR spectrum; 10% in CCl$_4$ solution (3800 to 1333 cm^{-1}) and 10% in CS$_2$ solution (1333 to 450 cm^{-1}). Solvents are not compensated. Bands at 658 and 667 cm^{-1} result from the presence of O,O-dimethyl phosphorochloridothioate as an impurity. Lower: Raman spectrum; neat liquid.

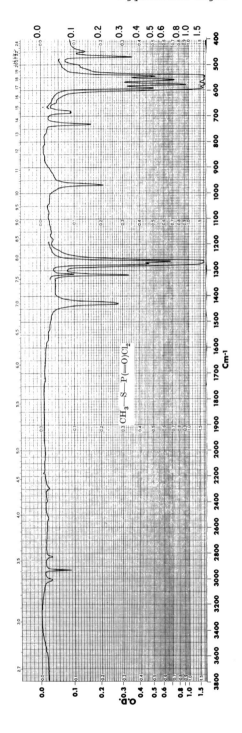

$CH_3-S-P(=O)Cl_2$

Cm⁻¹

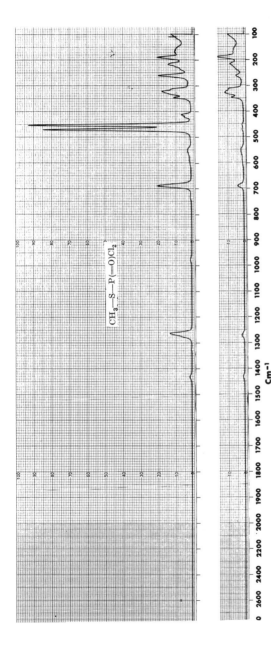

FIG. 45. S-Methyl phosphorodichloridothioate. Upper: IR spectrum; 10% in CCl$_4$ solution (3800 to 1333 cm^{-1}). Lower: Raman spectrum and polarized Raman spectrum; neat liquid.

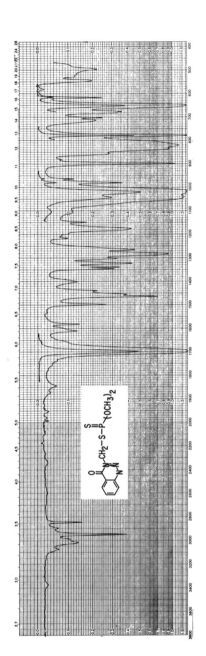

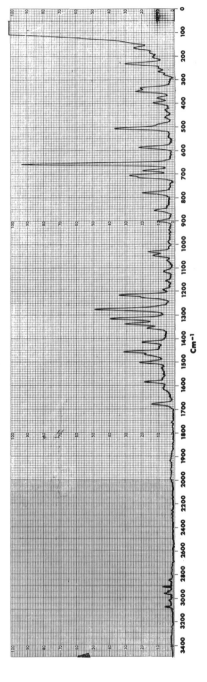

FIG. 46. O,O-Dimethyl S-[4-oxo-1,2,3-benzotriazin-3(4H)-yl methyl]phosphorodithioate. Upper: IR spectrum;
2 and 10% in CCl4 solution (3800 to 1333 cm⁻¹) and 2 and 10% in CS2 solution (1333 to 450 cm⁻¹). Solvents
are compensated. Lower: Raman spectrum; neat liquid.

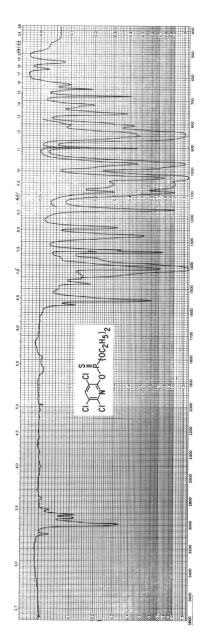

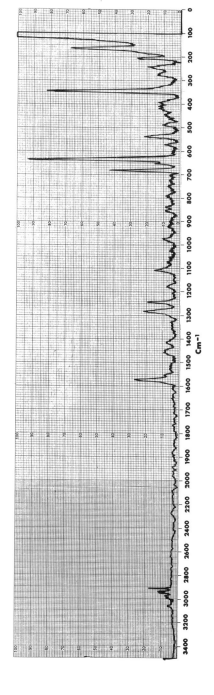

FIG. 47. O,O-Diethyl O-(3,5,6-trichloro-2-pyridyl)phosphorothioate. Upper: IR spectrum; 2 and 10% in CCl₄ solution (3800 to 1333 cm⁻¹) and 2 and 10% in CS₂ solution (1333 to 450 cm⁻¹). Solvents are compensated. Lower: Raman spectrum; neat solid.

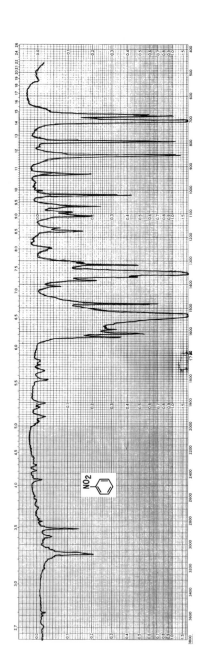

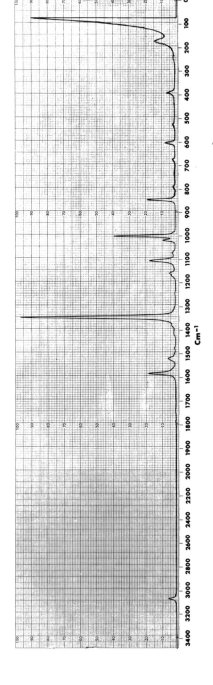

FIG. 48. Nitrobenzene. Upper: IR spectrum; 10% in CCl₄ solution (3800 to 1333 cm⁻¹) and 10% in CS₂ solution (1333 to 450 cm⁻¹). Lower: Raman spectrum; neat liquid.

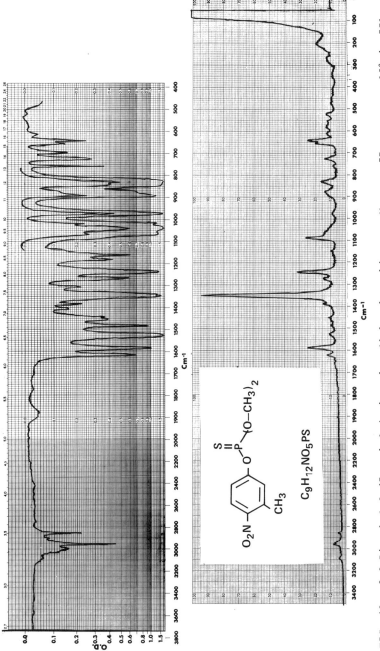

FIG. 49. O,O-Dimethyl O-(3-methyl-4-nitrophenyl)phosphorothioate. Upper: IR spectrum; 10% in CCl₄ solution (3800 to 1333 cm⁻¹) and 10 and 2% in CS₂ solution (1333 to 450 cm⁻¹). Solvents are compensated. Lower: Raman spectrum; neat liquid.

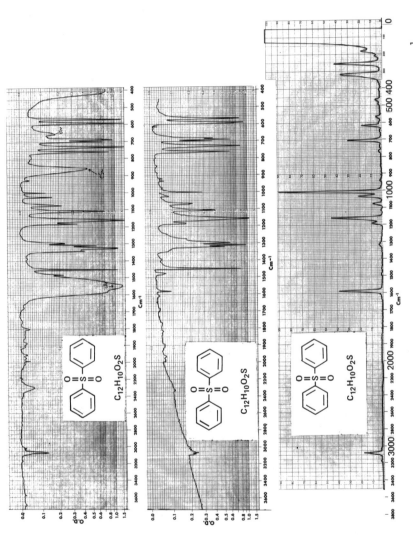

FIG. 50. Diphenyl sulfone. Upper: IR spectrum; 1% in CCl4 solution (3800 to 1333 cm^{-1}) and 1% in CS$_2$ solution (1333 to 450 cm^{-1}). Solvents are not compensated. Middle: IR spectrum; split mull [Fluorolube (3800 to 1333 cm^{-1}) and Nujol (1333 to 450 cm^{-1})]. Lower: Raman spectrum; neat solid.

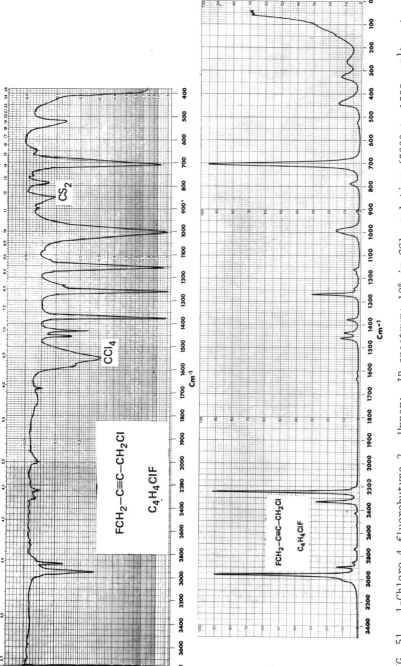

FIG. 51. 1-Chloro-4-fluorobutyne-2. Upper: IR spectrum; 10% in CCl₄ solution (3800 to 1333 cm⁻¹) and 10% in CS₂ solution (1333 to 450 cm⁻¹). Lower: Raman spectrum; neat liquid. Solvents are not compensated.

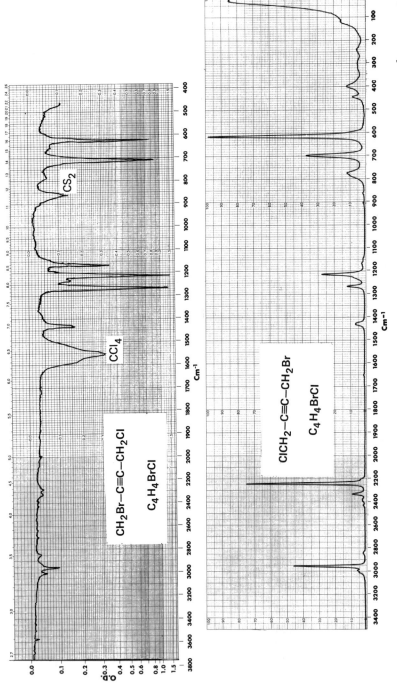

FIG. 52. 1-Bromo-4-chlorobutyne-2. Upper: IR spectrum; 10% in CCl₄ solution (3800 to 1333 cm⁻¹) and 10% in CS₂ solution (1333 to 450 cm⁻¹). Lower: Raman spectrum; neat liquid. Solvents are not compensated.

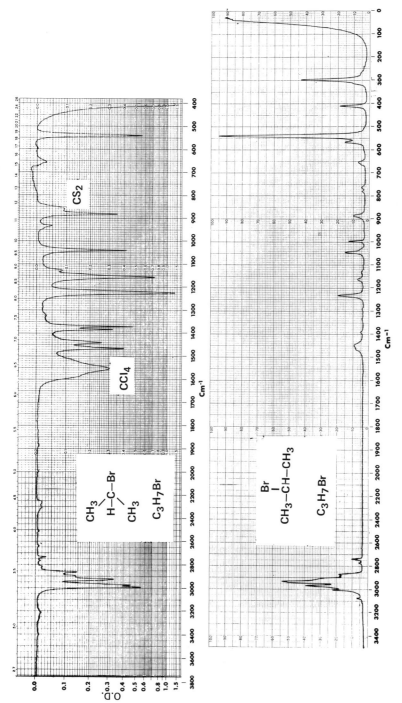

FIG. 53. 2-Bromopropane. Upper: IR spectrum; 10% in CCl$_4$ solution (3800 to 1333 cm^{-1}) and 10% in CS$_2$ solution (1333 to 450 cm^{-1}). Lower: Raman spectrum; neat liquid. Solvents are not compensated.

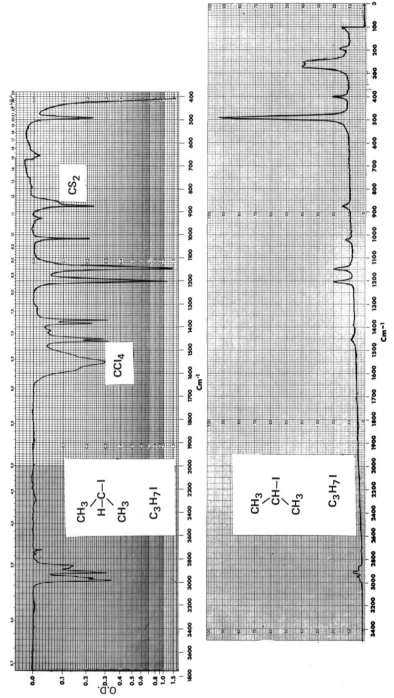

FIG. 54. 2-Iodopropane. Upper: IR spectrum; 10% in CCl$_4$ solution (3800 to 1333 cm^{-1}) and 10% in CS$_2$ solution (1333 to 450 cm^{-1}). Solvents are not compensated. Lower: Raman spectrum; neat liquid.

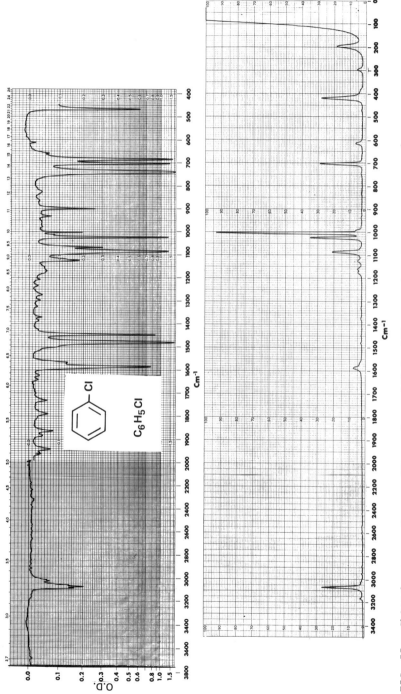

FIG. 55. Chlorobenzene. Upper: IR spectrum; 10% in CCl₄ solution (3800 to 1333 cm⁻¹) and 10% in CS₂ solution (1333 to 450 cm⁻¹). Solvents are compensated. Lower: Raman spectrum; neat liquid.

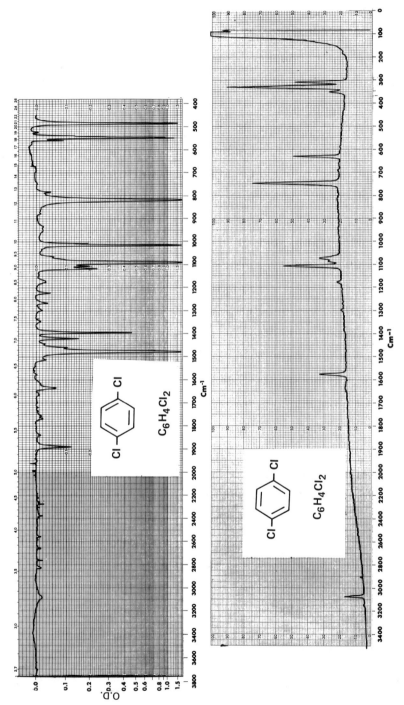

FIG. 56. p-Dichlorobenzene. Upper: IR spectrum; 10% in CCl$_4$ solution (3800 to 1333 cm^{-1}) and 10% in CS$_2$ solution (1333 to 450 cm^{-1}). Solvents are compensated. Lower: Raman spectrum; neat solid.

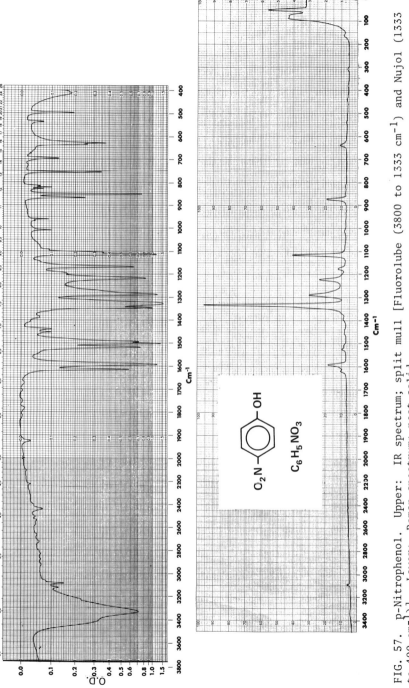

FIG. 57. p-Nitrophenol. Upper: IR spectrum; split mull [Fluorolube (3800 to 1333 cm^{-1}) and Nujol (1333 to 400 cm^{-1})]. Lower: Raman spectrum; neat solid.

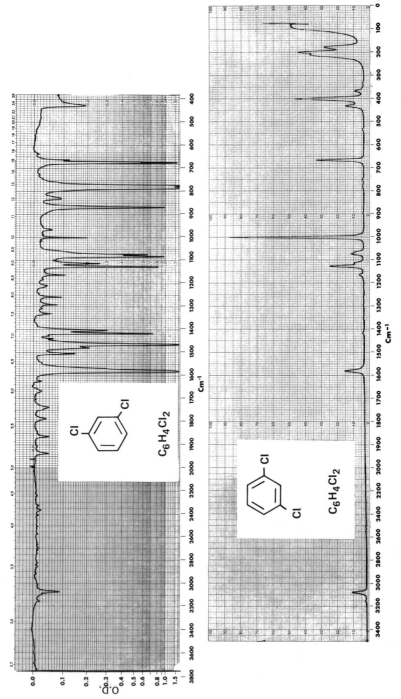

FIG. 58. m-Dichlorobenzene. Upper: IR spectrum; 10% in CCl$_4$ solution (3800 to 1333 cm^{-1}) and 10% in CS$_2$ solution (1333 to 400 cm^{-1}). Solvents are compensated. Lower: Raman spectrum; neat liquid.

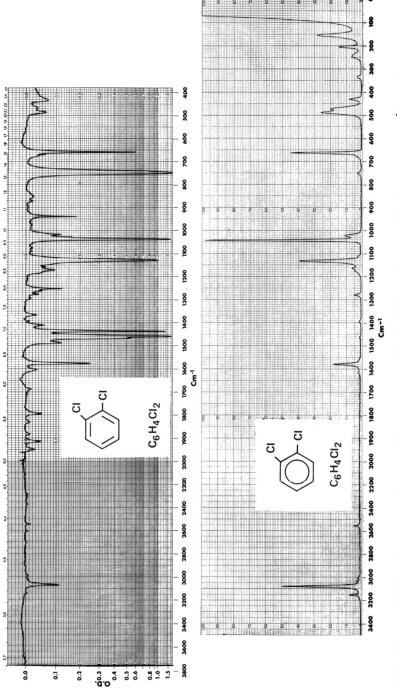

FIG. 59. o-Dichlorobenzene. Upper: IR spectrum; 10% in CCl$_4$ solution (3800 to 1333 cm^{-1}) and 10% in CS$_2$ solution (1333 to 400 cm^{-1}). Lower: Raman spectrum; neat liquid.

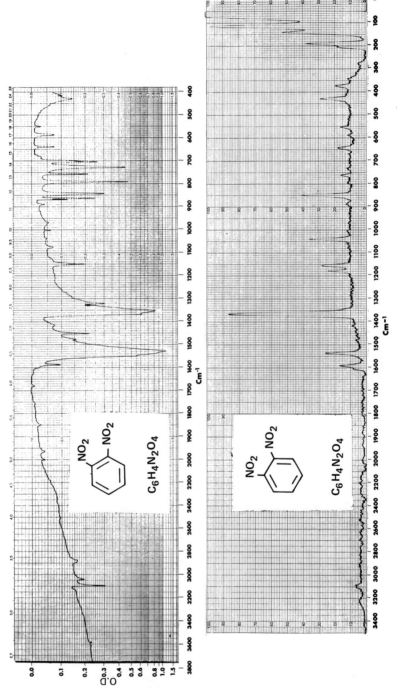

FIG. 60. o-Dinitrobenzene. Upper: IR spectrum; split mull [Fluorolube (3800 to 1333 cm⁻¹) and Nujol (1333 to 400 cm⁻¹)]. Lower: Raman spectrum; neat solid.

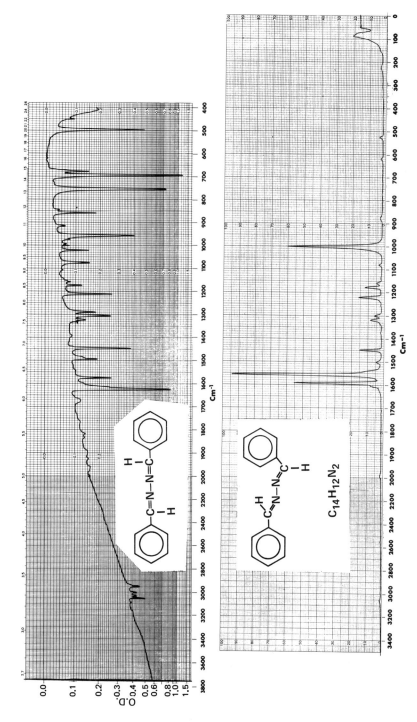

FIG. 61. Azine: benzaldehyde. Upper: IR spectrum; split mull [Flurolube (3800 to 1333 cm⁻¹) and Nujol (1333 to 400 cm⁻¹)]. Lower: Raman spectrum; neat solid.

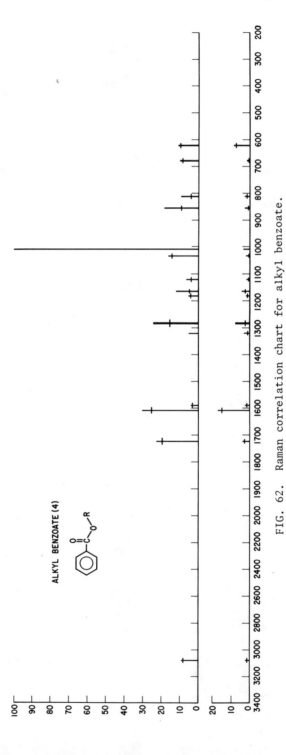

ALKYL BENZOATE (4)

FIG. 62. Raman correlation chart for alkyl benzoate.

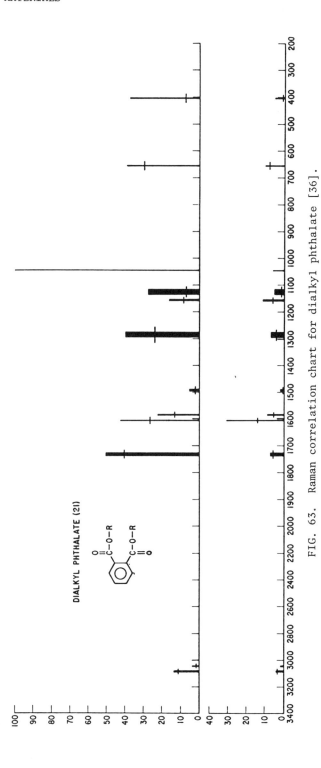

FIG. 63. Raman correlation chart for dialkyl phthalate [36].

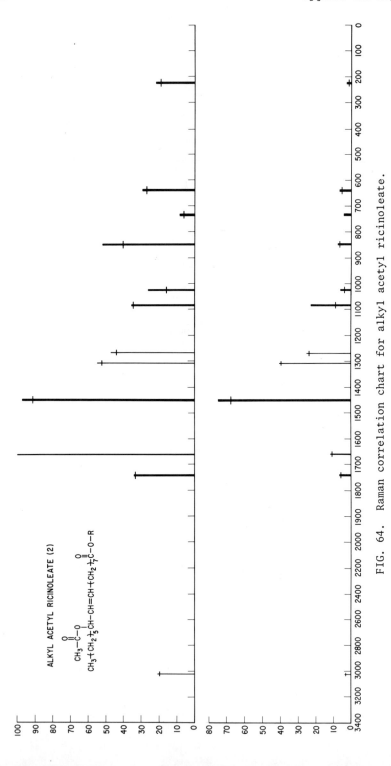

FIG. 64.   Raman correlation chart for alkyl acetyl ricinoleate.

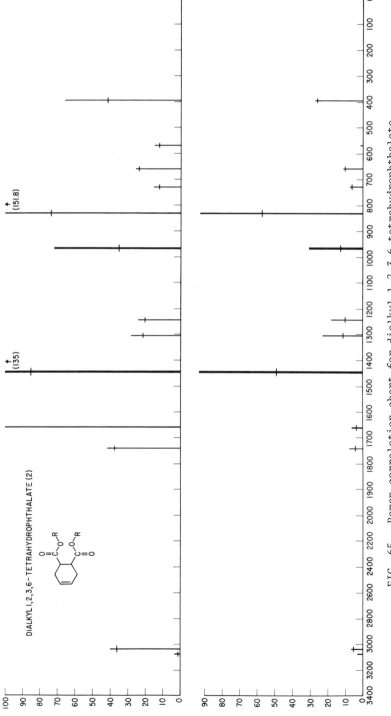

FIG. 65. Raman correlation chart for dialkyl 1,2,3,6-tetrahydrophthalate.

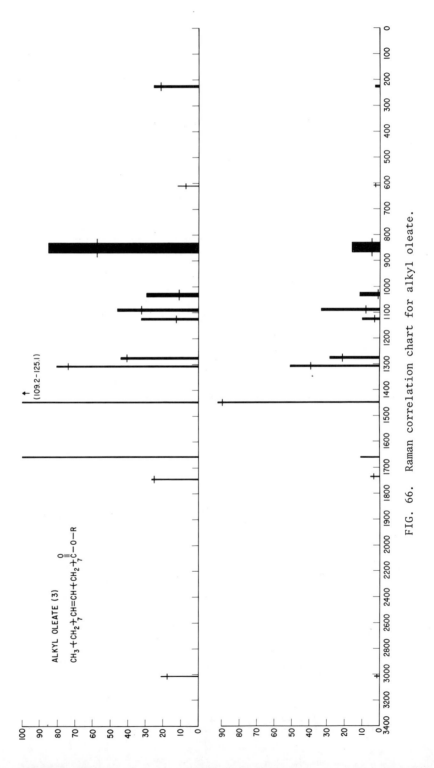

FIG. 66. Raman correlation chart for alkyl oleate.

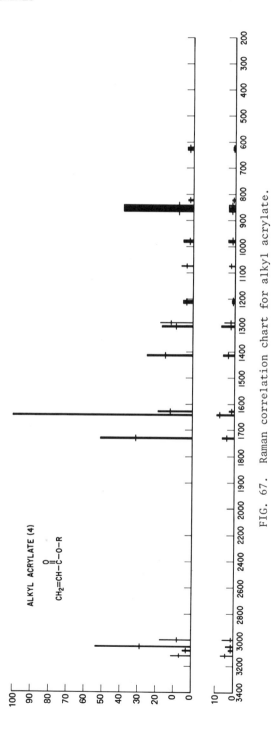

FIG. 67. Raman correlation chart for alkyl acrylate.

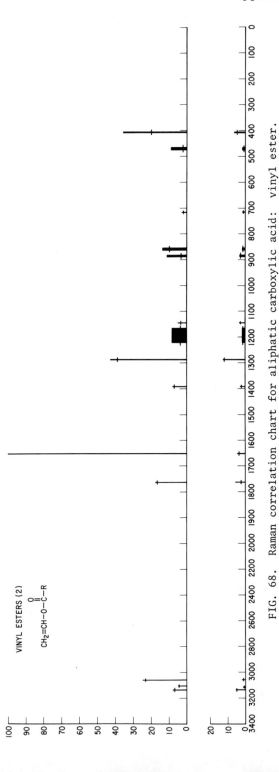

FIG. 68. Raman correlation chart for aliphatic carboxylic acid: vinyl ester.

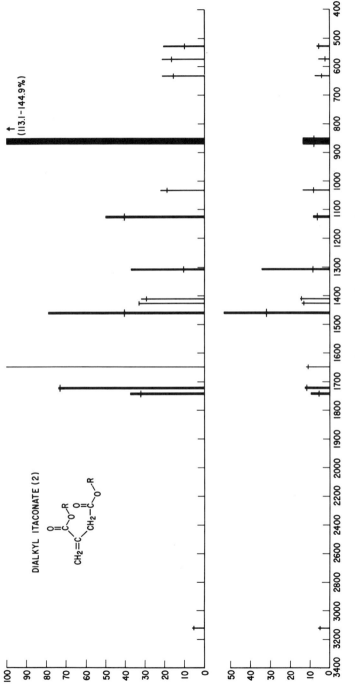

FIG. 69.   Raman correlation chart for dialkyl itaconate.

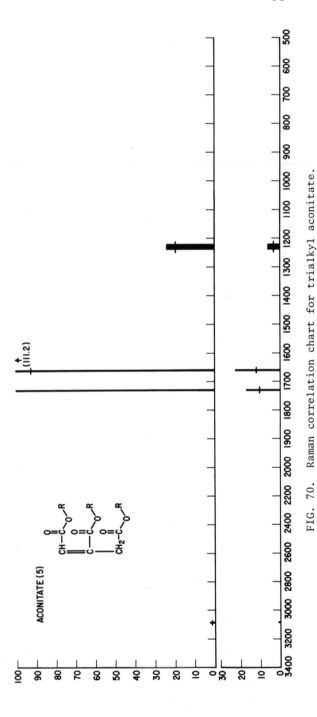

FIG. 70. Raman correlation chart for trialkyl aconitate.

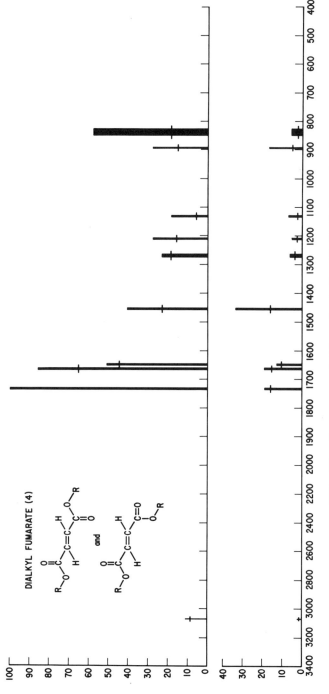

FIG. 71. Raman correlation chart for dialkyl fumarate.

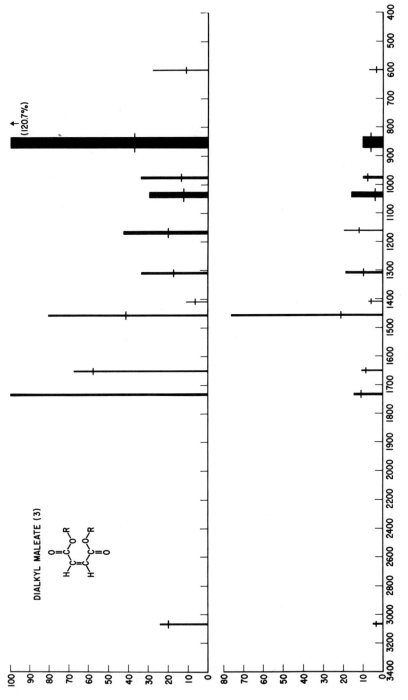

FIG. 72.   Raman correlation chart for dialkyl maleate.

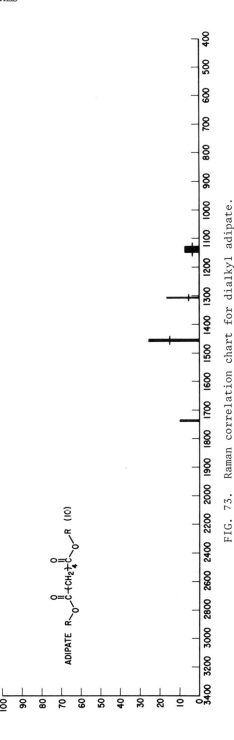

FIG. 73. Raman correlation chart for dialkyl adipate.

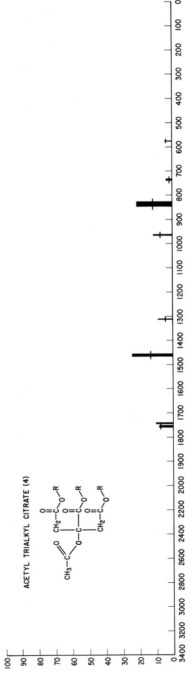

FIG. 74.   Raman correlation chart for acetyl trialkyl citrate.

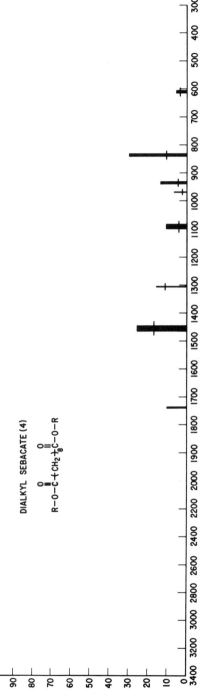

FIG. 75. Raman correlation chart for dialkyl sebacate.

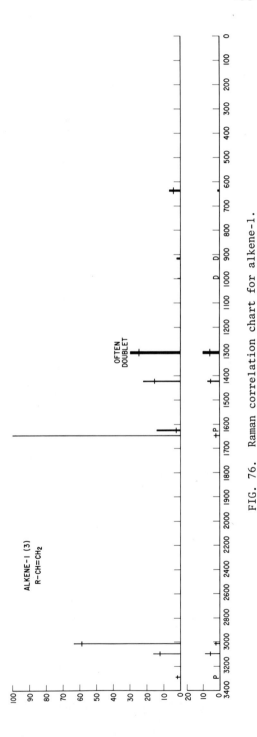

FIG. 76.   Raman correlation chart for alkene-1.

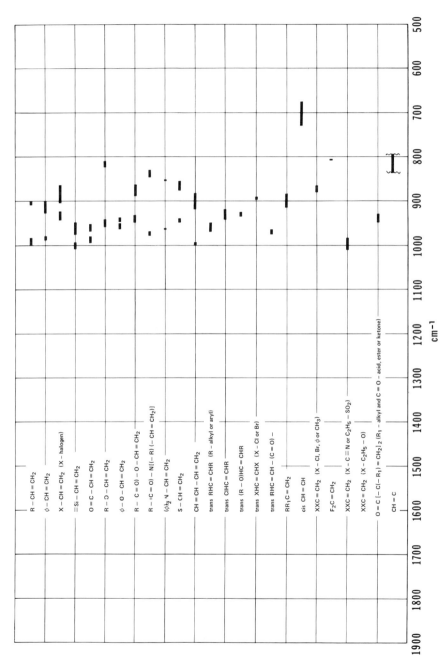

FIG. 77.   Infrared correlation chart for out-of-plane olefinic hydrogen deformations.

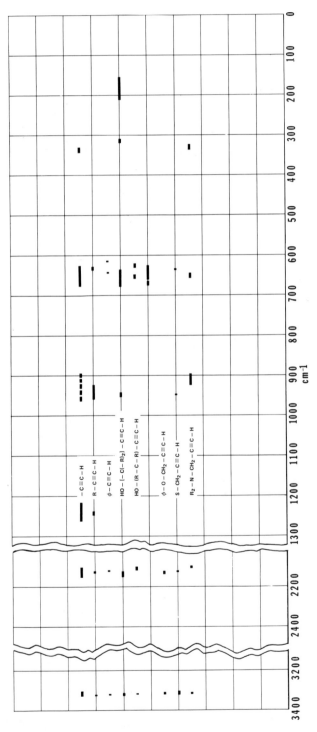

FIG. 78.   Infrared correlation chart for terminal acetylenic group frequencies.

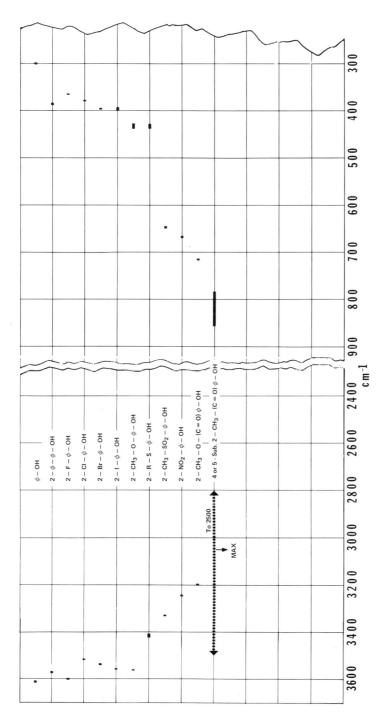

FIG. 79. Infrared correlation chart for hydroxyl stretching and out-of-plane bending vibrations of intra-molecularly hydrogen-bonded phenols.

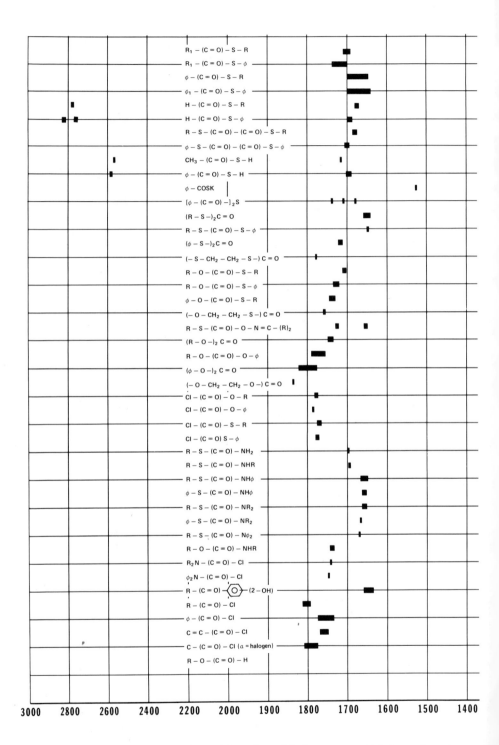

FIG. 80.   Infrared correlation chart for thiol esters, thiol carbonates,

554

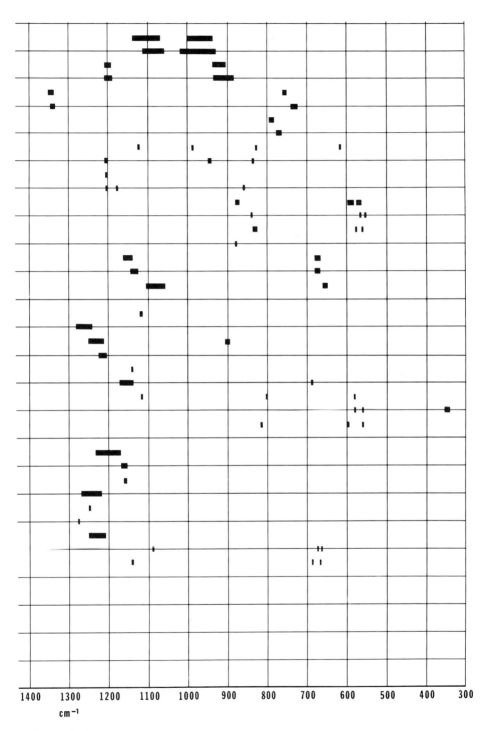

| 1400 | 1300 | 1200 | 1100 | 1000 | 900 | 800 | 700 | 600 | 500 | 400 | 300 |

cm⁻¹

and related compounds.

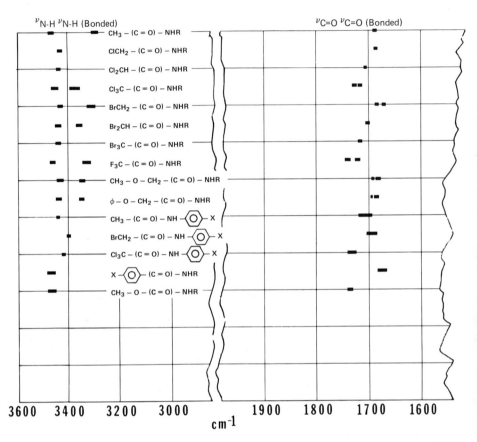

FIG. 81. Infrared correlation chart for amides and carbamates: N-H
and C=O stretching.

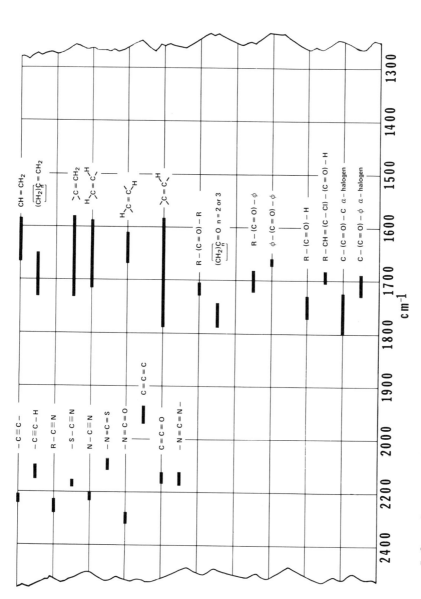

Infrared correlation chart for miscellaneous double and triple bond stretching vibrations.

Figure 82

REFERENCES

1. G. Herzberg, *Molecular Spectra and Molecular Structure,* Part II, Van Nostrand-Reinhold, Princeton, New Jersey, 1945.

2. E. B. Wilson, J. C. Decius, and P. C. Cross, *Molecular Vibrations,* McGraw-Hill, New York, 1955.

3. A. B. F. Duncan, *Chemical Applications of Spectroscopy,* Vol. IX, (W. West, ed.), Chap. III, Wiley (Interscience), New York, 1956.

4. N. B. Colthup, L. H. Daly, and S. E. Wiberly, *Introduction to Infrared and Raman Spectroscopy,* Academic Press, New York, 1964.

5. W. J. Potts, Jr., *Chemical Infrared Spectroscopy,* Wiley, New York, 1963.

6. R. G. White, *Handbook of Industrial Infrared Analysis,* Plenum Press, New York, 1964.

7. C. N. R. Rao, *Chemical Applications of Infrared Spectroscopy,* Academic Press, New York, 1963.

8. R. A. Nyquist and R. O. Kagel, *Infrared Spectra of Inorganic Compounds (3800-45 cm^{-1}),* Academic Press, New York, 1971.

9. R. D. McLachlan, private communication.

10. R. A. Nyquist and W. W. Muelder, *Spectrochim. Acta, 22,* 1563 (1966).

11. R. N. Jones and C. Sandorfy, *Chemical Applications of Spectroscopy,* Vol. IX, (W. West, ed.), Chap. IV, Wiley (Interscience), New York, 1956.

12. S. T. King, *J. Agr. and Food Chem., 21,* 529 (1973).

13. S. T. King, 2nd Fourier Transform Users' Group Meeting, Pittsburg Conference on Analytical Chemistry and Applied Spectroscopy, Cleveland, Ohio, 1973.

14. P. R. Griffith, 2nd Fourier Transform Users' Group Meeting, Pittsburg Conference on Analytical Chemistry and Applied Spectroscopy, Cleveland, Ohio, 1973.

15. G. F. Bailey, S. Kint, and J. R. Scherer, *Anal. Chem., 39,* 1040 (1967).

16. S. K. Freeman, P. R. Reed, and D. O. Landon, *Mikrochim Acta,* 288, 1972.

17. R. O. Kagel and R. A. Nyquist, Paper 284, Abstracts of Pittsburg Conference on Analytical Chemistry and Applied Spectroscopy, Cleveland, Ohio, 1972.

18. B. J. Bulkin, K. Dill, and J. J. Dannenberg, *Anal. Chem., 43,* 974 (1971).

19. P. R. Griffith, *Appl. Spectr., 26,* 73 (1972).

20. M. J. D. Low and H. Mark, *J. Paint Technol., 42,* 265 (1970).

21. K. H. Rhee and L. R. Cousins, *Spectrometry of Fuels, 26,* 1633, 1970.

22. R. O. Kagel, 1st Fourier Transform Users' Group Meeting, Eastern Analytical Symposium, Atlantic City, New Jersey, 1972.

23. T. R. Kozlowski, *Appl. Opt., 7,* 795 (1968).

24. C. W. Brown and S. F. Baldwin, *Water Res., 6,* 1601 (1972).

25. L. Rimai, R. G. Kilponen, and D. Gill, *Biochem and Biophys. Res. Comm., 41,* 492 (1970).

26. L. J. Bellamy, *The Infrared Spectra of Complex Molecules,* (John Wiley and Sons, Inc.), New York, 1958.

27. L. J. Bellamy, *Advances in Infrared Group Frequencies,* Methuen, London, 1968.

28. R. A. Nyquist and W. J. Potts, Jr. in *Analytical Chemistry of Phosphorus Compounds* (M. Halmann, ed.), Chap. 5, Wiley, New York, 1972.

29. N. Sheppard, *Trans. Faraday Soc., 46,* 429 (1950).

30. D. W. Scott, H. L. Finke, W. N. Hublard, J. P. McCullough, G. D. Oliver, M. E. Gross, C. Katz, K. D. Williamson, S. Waddington, and H. M. Huffman, *J. Amer. Chem. Soc., 74,* 4656 (1952).

31. R. A. Nyquist, *Spectrochim. Acta,* 23A, *845* (1967).

32. R. A. Nyquist, *Spectrochim. Acta,* 23A, 2505 (1967).

33. R. A. Nyquist, *Spectrochim. Acta, 27A,* 697 (1971).

34. J. R. Scherer, *Planar Vibrations of Chlorinated Benzenes* (The Dow Chemical Co., 1960).

35. J. R. Scherer, *Spectrochim. Acta, 19,* 601 (1963).

36. J. R. Scherer, *Spectrochim. Acta, 21,* 321 (1965).

37. R. A. Nyquist in *CRC Handbook of Spectroscopy-Infrared Spectroscopy,* (A. L. Smith, ed.), Chap. IV, Ohio, 1973.

38. R. A. Nyquist, *Appl. Spectr., 26,* 81 (1972).

39. W. J. Potts, Jr., and R. A. Nyquist, *Spectrochim. Acta, 7,* 514 (1959).

40. R. A. Nyquist and W. J. Potts, Jr., *Spectrochim. Acta, 16,* 419 (1960).

41. R. A. Nyquist, *Spectrochim. Acta, 19,* 1655 (1963).

42. R. A. Nyquist and W. J. Potts, Jr., *Spectrochim. Acta, 7,* 514 (1959).

43. R. A. Nyquist and W. J. Potts, Jr., *Spectrochim. Acta, 17,* 679 (1961).

44. R. A. Nyquist, *Spectrochim. Acta, 19,* 508 (1963).

45. R. D. McLachlan and R. A. Nyquist, *Spectrochim. Acta, 19,* 1595 (1963).

APPENDIX: INDEXES OF SPECTRA AND CHARTS

A. Infrared Spectra--Alphabetical

	Fig.		Fig.
Acetic acid	19	3,5-Dinitro-o-toluamide	1
Acetyl salicylic acid	21	Diphenyl butadiyne	38
Acrylonitrile	2	Diphenyl sulfide	30
Allyl mercaptan	25	Diphenyl sulfone	50
Aspirin	21	DURSBAN (trademark of the Dow Chemical Company)	47
Azine: benzaldehyde	61	Ethyl fumarate	37
Benzaldehyde	24	Fumaronitrile	40
Benzoic acid	20		
Benzyl alcohol	7	Group frequency correlations--organophosphorus compounds	38
Benzoyl chloride	23	Guthion	42
1-Bromo-4-chlorobutyne-2	52	Insect	5
2-Bromopropane	53	2-Iodopropane	54
C-S stretching in mercaptans (frequency data)	27,29	Isopropyl bromide	53
		Isopropyl iodide	54
C-S stretching in sulfides (frequency data)	29	Maleonitrile	41
Chlorobenzene	55	Methyl acrylate	36
1-Chloro-4-fluorobutyne-2	51	O-Methyl phosphorodichlorido-thioate	44
p-Dichlorobenzene	56	S-Methyl phosphorodichlorido-thioate	45
m-Dichlorobenzene	58		
o-Dichlorobenzene	59	N-Methylphthalimide	18
1,2-Dichloroethylene (trans)	33	Nitrobenzene	48
Diethyl phthalate	17	1-Octene	35
0,0-Diethyl O-(3,5,6-trichloro-2-pyridyl)phosphorothioate	47	p-Nitrophenol	57
		Perchloroethylene	32
0,0-Dimethyl O-(3-methyl-4-nitrophenyl)phosphorothioate	49	Poly(ethylene)	12
0,0-Dimethyl S-[4-oxo-1,2,3-benzotriazin-3(4H)-yl methyl]phosphorothioate	46	S-H Stretching (frequency data)	26
		Sodium acetate	22
Dimethyl sulfide	28	S-S Stretching in disulfides (frequency data)	31
o-Dinitrobenzene	60		

A. Infrared Spectra--Alphabetical (continued)

	Fig.		Fig.
Sumithion	45	Thiophosphoryl dichloride fluoride	43
2,4,4,4-Tetrachlorobutyro-nitrile	39	Vinyl chloride	34
Tetrachloroethylene	32	Vinyl cyanide	2
		Zoalene	1

B. Infrared Correlation Charts

	Fig.		Fig.
Acetylenic (-C≡C-H)	78	Olefins (out-of-plane hydrogen deformations)	77
Amides	81	Organophosphorus compounds	72
Carbamates	81	Phenols (intramolecular hydrogen bonded)	79
Carbonates	80		
Double bonds	82	Thiol carbonates	80
Hydroxyl (intramolecular hydrogen bonded)	79	Thiol esters	80
		Triple bonds	78, 82

C. Infrared Spectra--Empirical Formula

	Fig.		Fig.
CH_3Cl_2OPS	44, 46	C_4H_4BrCl	52
C_2Cl_4	32	C_4H_4ClF	51
$C_2H_2Cl_2$	33	$C_4H_6O_2$	36
C_2H_3Cl	34	$C_6H_4Cl_2$	56, 58, 59
$C_2H_3NaO_2$	22	$C_6H_4N_2O_4$	60
$(C_2H_4)_n$	12	C_6H_5Cl	55
$C_2H_4O_2$	19	$C_6H_5NO_2$	48
C_2H_6S	28	$C_6H_5NO_3$	57
C_3H_3N	2	C_7H_5ClO	23
C_3H_6S	25	C_7H_6O	24
$C_4H_2N_2$	40, 41	$C_7H_6O_2$	20
$C_4H_3Cl_4N$	39	$C_8H_7N_3O_5$	1

C. Infrared Spectra--Empirical Formula (continued)

	Fig.		Fig.
$C_8H_{12}O_4$	37	$C_{10}H_{12}N_3O_3PS_2$	46
C_8H_{16}	35	$C_{12}H_{10}O_2S$	50
$C_9H_7NO_2$	18	$C_{12}H_{10}S$	30
$C_9H_8O_4$	21	$C_{12}H_{14}O_4$	17
$C_9H_{11}Cl_3NO_3PS$	47	$C_{14}H_{12}N_2$	61
$C_9H_{12}NO_5PS$	49	$C_{16}H_{10}$	38
		Cl_2FPS	43

D. Raman Correlation Charts

	Fig.		Fig.
Acetyl citric acid: trialkyl ester	74	Benzoic acid: alkyl ester	62
Acetyl ricinoleate: alkyl ester	64	Fumaric acid: dialkyl ester	72
Aconitic acid: dialkyl ester	70	Itaconic acid: dialkyl ester	69
Acrylic acid: alkyl ester	73	Maleic acid: dialkyl ester	72
		Oleic acid: alkyl ester	66
Aliphatic carboxylic acid: vinyl ester	68	Phthalic acid: dialkyl ester	63
		Sebacid acid: dialkyl ester	75
Alkene-1	76	1,2,3,6-Tetrahydrophthalic acid: dialkyl ester	65

E. Raman Spectra--Alphabetical

	Fig.		Fig.
Acetic acid	19	Benzyl alcohol	8, 16
Acetyl salicylic acid	21	Benzoyl chloride	23
Acrylonitrile	2, 4	1-Bromo-4-chlorobutyne-2	52
Allyl mercaptan	25	2-Bromopropane	53
Aspirin	21	C-S Stretching in mercaptans (frequency data)	27, 29
Azine: benzaldehyde	61		
Benzaldehyde	24	C-S Stretching in sulfides (frequency data)	29
Benzoic acid	20	Chlorobenzene	55

E. Raman Spectra--Alphabetical (continued)

	Fig.		Fig.
1-Chloro-4-fluorobutyne-2	51	Isopropyl iodide	54
p-Dichlorobenzene	56	Maleonitrile	41
m-Dichlorobenzene	58	Methyl acrylate	36
o-Dichlorobenzene	59	O-Methyl phosphorodichlorido-thioate	44
1,2-Dichloroethylene (trans)	33		
Diethyl fumarate	37	S-Methyl phosphorodichlorido-thioate	45
Diethyl phthalate	17	N-Methylphthalimide	18
O,O-Diethyl O-(3,4,5-trichloro-2-pyridyl)phosphorothioate	47	Nitrobenzene	48
O,O-Dimethyl O-(3-methyl-4-nitrophenyl)phosphorothioate	49	1-Octene	35
		p-Nitrophenol	57
O,O-Dimethyl S-[4-oxo-1,2,3-benzotriazin-3(4H)-yl methyl]phosphorothioate	46	Perchloroethylene	15, 32
		S-H Stretching (frequency data)	26
Dimethyl sulfide	28	Sodium acetate	22
o-Dinitrobenzene	60	S-S Stretching in disulfides (frequency data)	31
Diphenyl butadiyne	38		
Diphenyl sulfide	30	Sumithion	49
Diphenyl sulfone	50	2,4,4,4-Tetrachlorobutyro-nitrile	39
DURSBAN (trademark of the Dow Chemical Company)	47	Tetrachloroethylene	32
Ethyl fumarate	37	1,2,2-Tribromopropane	10
Fumaronitrile	40	1,2,3-Tribromopropane	10
Guthion	46	Thiophosphoryl dichloride fluoride	43
2-Iodopropane	54	Vinyl chloride	34
Isopropyl bromide	53	Vinyl cyanide	2

F. Raman Spectra--Empirical Formula

	Fig.		Fig.
CH_3Cl_2OPS	44, 45	C_2H_3Cl	34
C_2Cl_4	32	$_2H_3NaO_2$	22
$C_2H_2Cl_2$	33	$C_2H_4O_2$	19

F. Raman Spectra--Empirical Formula (continued)

	Fig.		Fig.
C_2H_6S	28	$C_7H_6O_2$	20
C_3H_3N	2	C_7H_8O	8, 16
C_3H_6S	25	$C_8H_7N_3O_5$	1
$C_4H_2N_2$	40, 41	$C_8H_{12}O_4$	37
$C_4H_3Cl_4N$	39	C_8H_{16}	35
C_4H_4BrCl	52	$C_9H_7NO_2$	18
C_4H_4ClF	51	$C_9H_8O_4$	21
$C_4H_6O_2$	36	$C_9H_{11}Cl_3NO_3PS$	47
$C_6H_4Cl_2$	56, 58, 59	$C_9H_{12}NO_5PS$	49
$C_6H_4N_2O_4$	60	$C_{10}H_{12}N_3O_3PS_2$	46
C_6H_5Cl	55	$C_{12}H_{10}O_2S$	50
$C_6H_5NO_2$	48	$C_{12}H_{10}S$	30
$C_6H_5NO_3$	57	$C_{12}H_{14}O_4$	17
C_7H_5ClO	23	$C_{14}H_{12}N_2$	61
C_7H_6O	24	$C_{16}H_{10}$	38
		Cl_2FPS	43

Chapter 7

ENVIRONMENTAL SCIENCE

Donald S. Lavery*

Marketing Department
Wilks Scientific Corporation
South Norwalk, Connecticut

I.   INTRODUCTION 565
II. GAS ANALYSIS 567
     A. Special Equipment for Gas Analysis 567
     B. Minimum Detection Limits of Infrared Methods 574
     C. Calibration Methods for Gas Analysis 575
     D. Sample Handling. 598
     E. Analysis of Multicomponent Mixtures. 609
     F. Remote Sensing Methods 611
III. WATER ANALYSIS 615
     A. Dissolved Organic Contaminants 615
     B. Floating Films .618
     C. Inorganic Ions .621

     REFERENCES 621

I. INTRODUCTION

The environmental applications of vibrational spectroscopy cover a
wide spectrum of activities which range in scope from routine analytical measurements using simple equipment to exotic research projects

---

*Present affiliation: Consultant in Analytical Chemistry, East
Norwalk, Connecticut.

using elaborate, specialized equipment. They encompass fields of
scientific interest from physics and physical chemistry to biochemistry and biology. Needless to say, no attempt is made in this chapter to make this treatment comprehensive. Rather, most of the discussion and virtually all of the data are concerned with the kinds
of analytical measurements which can be made with ordinary commercial
instruments. Moreover, the discussion has been confined to the area
of analytical spectroscopy, and no attempt was made to assess the
significance of the environmental problems which prompt the analytical studies.

Most of the present discussion concerns infrared (IR) methods
for several reasons:

1. Despite the recent tremendous growth in Raman spectroscopy
since the introduction of commercial laser sources, IR equipment is
still much more common than Raman equipment.

2. Gas analysis, which is probably the most important use of
vibrational spectroscopy in environmental problems, is much simpler
by IR than by Raman methods.

3. The large amount of IR spectral data available in the literature makes the application of IR to new problems much more straightforward than less common techniques.

Some discussion of specialized studies which are entirely beyond
the reach of the average analytical laboratory is given in order to
give a somewhat more balanced view of the field as a whole and to
indicate likely directions of future developments. Readers are referred to other sources for detailed discussions of these studies
which include research in atmospheric physics and chemistry, remote
sensing by Raman and IR laser methods, ultrahigh resolution spectroscopy, and correlation methods.

II. GAS ANALYSIS

A. Special Equipment For Gas Analysis

Most of the gas analyses required for environmental studies in-volve gases at low concentration and consequently require the use of extended path cells. Such cells have been in use for many years but have not, until recently, been common in analytical laboratories. The rather detailed discussion which follows therefore seems warranted.

1. Long Path Gas Cells

A number of two-, three-, and four-pass cells with pathlengths up to about 1 m are available. They are described elsewhere [2]. Cells in this pathlength range are useful for studies of samples like auto exhaust, stack gas from combustion processes, duct samples from solvent evaporation processes, and for indoor air analysis when moderately strongly-absorbing materials are involved. They do not usually provide sufficient sensitivity for direct analysis of out-door air. Efficient long path cells have been built with optical pathlengths up to about 1 km. Cells with optical paths in the 20-m and 40-m range are now available for use with almost all commercial spectrometers and are becoming quite common.

a. *The White Cell.* The first efficient multipass cell was designed by J. White [1]. The most important feature introduced in this design was the use of a field mirror. This point is still not appreciated widely by IR spectroscopists. The three-mirror system used by White is shown schematically in Fig. 1, but for our prelimin-ary discussion it is convenient to consider the equivalent lens sys-tem shown in Fig. 2. Figure 2(a) is equivalent to a three-mirror cell such as that of Fig. 1 but having a plane mirror at the left-hand end. Figure 2(b) is completely equivalent to the cell of Fig. 1. The objective lenses O_1 and O_2 are separated by four times their focal length. If we consider an extended source S, rays A and B

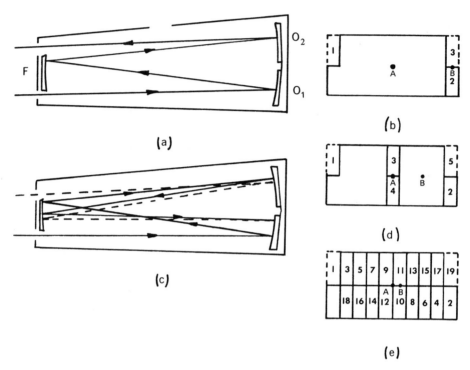

FIG. 1. White cell of physical length ℓ: (a) ray diagram for minimum pathlength 4ℓ; (b) image pattern at field mirror for pathlength
4ℓ; (c) ray diagram for pathlength 4ℓ; (d) image pattern for 8ℓ;
(e) image pattern for 36ℓ (A and B are centers of curvature of O_1 and
O_2, respectively).

[Fig. 2(a)] serve to locate the real image I_1, formed by O_1. These
rays are then collected by O_2 and form an image at I_2. Consideration
of only these rays would seem to suggest that this process could be
continued for several stages and that such a system would serve as a
long path cell. Ray C shows, however, that rays properly focused at
I_1 can miss O_2 entirely and be lost. This loss constitutes a reduction in the aperture of the system and after several reimaging stages
this reduction can be so severe that the system will be incapable of
transmitting useful amounts of energy. White's solution to this
problem is illustrated in Fig. 2(b). Lens F, which images O_1 onto
O_2, ensures that all rays from O_1 will fall on O_2, eliminating the

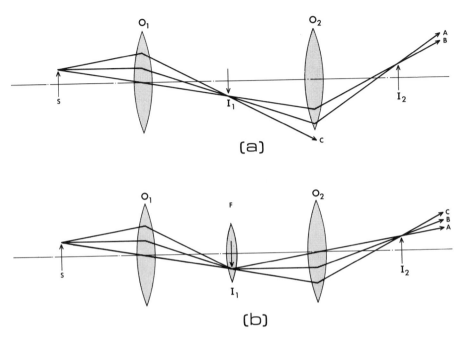

FIG. 2. Lens system equivalent to long path cells. (a) Vignetting in an extended optical system; (b) elimination of vignetting by a field lens.

problem of aperture reduction. Figure 1(a) shows the actual path of light through the cell when it is adjusted to its minimum pathlength. For a cell with a physical length of 1/2 m this would be a pathlength of 2 m. Only the central ray is shown but in fact the beam diverges to fill mirror O_1, is focused to an image in the plane of F, diverges to fill O_2, and is focused again at F. Figure 1(b) shows the corresponding pattern of images formed in the plane of the field mirror. These field mirror diagrams are very useful in following the light beam through the cell for various pathlengths. Figures 1(c) and (d) show the next longer pathlength (4 m), while Fig. 1(e) shows the maximum pathlength (18 m) with the images just touching. Longer pathlengths can only be obtained by reducing the size of the entrance aperture (which is excessively large in this diagram). In practice, long path cells are usually designed so that the practical

pathlength limit is determined by reflection and aberration losses
of the optical system. The following analysis shows how this limit
arises when only reflection losses are important.

When light is reflected from a mirror, it is reduced in intensity by losses in the process. The reflectivity R of the surface is
the fraction of the incident light which is reflected. The loss on
reflection is then $(1 - R)$. After η reflections the fraction of
light reflected is R^{η} and the loss $(1 - R^{\eta})$. As is illustrated in
Fig. 1, increasing the path from any path to the next longer path
adds 4ℓ (ℓ = physical length of the cell) to the pathlength and adds
four reflections. The total pathlength will now be $(\eta + 1)\ell$, where
η is the total number of reflections at the field mirror and the
objective mirrors. Then,

$$\text{Fractional increase in path} = \frac{4\ell}{(\eta + 1)\ell} \tag{1}$$

$$\text{Fractional intensity loss} = 1 - R^4 \tag{2}$$

At short pathlengths the increase in pathlength clearly exceeds the
energy loss. At some point, however, the pathlength increase will
equal the energy loss. At pathlengths longer than this, the energy
loss will exceed the pathlength gain. At this point, we actually
lose sensitivity by further increasing pathlength.* The point at
which fractional energy loss equals fractional pathlength gain is
the maximum useful pathlength. Equating (1) and (2),

$$\frac{4\ell}{(\eta + 1)\ell} = 1 - R^4$$

$$\eta = \frac{4}{1 - R^4} - 1 \tag{3}$$

Typical metal reflectivities are 0.95 for evaporated aluminum and
0.985 for evaporated gold. Table 1 gives some calculated data based
on (3) for three different reflectivities. Cell transmission at the
maximum useful pathlength is about 35% exclusive of window losses;
including window losses would make this about 30%. Performance of
an actual cell is discussed in the next section.

––––––––––––––

*Assuming the sensitivity is limited by instrument noise.

TABLE 1

Maximum Useful Pathlength for Multipass Gas Cells
as a Function of Mirror Reflectivity

Reflectivity	Maximum number of reflections	Maximum useful pathlength	
		3/8 m Physical length	1 m Physical length
0.95	21	8.25	22
0.98	51	19.5	52
0.99	101	38.25	102

The method of changing pathlengths with this type of cell is indicated by the diagrams of Fig. 1. The points A and B are the centers of curvature of the objective mirrors O_1 and O_2. To increase pathlength these centers must be moved closer together as shown in Figs. 1(b), 1(d), 1(e). The optical path within the cell is conveniently traced using the field mirror diagrams of Figs. 1(b), 1(d), 1(e). Recall that a source in the plane of the center of curvature of a concave mirror will be imaged in that plane with the image inverted across the axis from the source. In Fig. 1(b), the entrance window (labeled 1) is the source and its image is formed at 2, inverted through A which lies on the axis of O_1. Since the light arriving at 2 is reflected by the field mirror, this becomes a source of light proceeding to O_2 which forms an image (inverted through B) at 3, the exit window, and the beam leaves the cell. The more complicated cases can be traced similarly. Changing the pathlength in a White cell requires that the objective mirrors be rotated toward each other. As long pathlengths are approached, the amount that the center B must move (i.e., the angular adjustment of O_2), to change from one pathlength to the next becomes very small. This makes adjustment difficult and makes the cell very sensitive to vibration and misalignment. A mechanically superior arrangement is described in the next section.

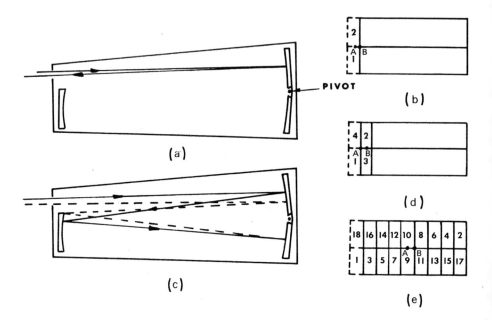

FIG. 3. Gilby cell of physical length ℓ: (a) ray diagram for pathlength 2ℓ; (b) image pattern at field mirror for pathlength 2ℓ;
(c) ray diagram for pathlength 6ℓ; (d) image pattern for 6ℓ; (e) image
pattern for 34ℓ (A and B are centers of curvature of the objective
mirrors).

    b. The Gilby Cell. Figure 3 shows a three-mirror multipass
cell which solves the aperture problem in the same way as does the
White cell but differs in the image arrangement on the field mirror.
Figures 3(a) and 3(b) show the minimum pathlength, Figs. 3(c) and
3(d) the next longest pathlength, and Fig. 3(e) the image arrangement
for the longest pathlength. As these figures show, pathlength is
changed by moving the centers of curvature of the objective mirrors
without changing their separation. The required motion is obtained
by rotating the two mirrors and their mounting plate about the pivot
point shown in the figure. The mechanical simplicity of this arrangement has allowed cells to be constructed which are rugged enough that
optical alignment is essentially permanent. The system has several
other advantages over the White cell. It is slightly more compact
and is less astigmatic at short pathlengths.

In deciding what the maximum useful path actually is for multi-pass cells, the considerations of the previous section may be applied. Instead of considering only reflection losses, the actual cell performance can be used, i.e., losses from reflection, scattering, aberrations, etc., will all be included. Table 2 shows the actual performance data for a Gilby cell with a physical length of 3/8 m. The average loss per step is 0.069. From this and Eq. (2), we can calculate the effective reflectivity to be 0.984. From Eq. (3), η = 57 and the maximum useful path is 21.75 m.

c. *The Hanst Cell.* Hanst [2] described an eight-mirror long path cell which is similar to the previous cells in having field

TABLE 2

Performance of a Multipass Cell[a]

Pathlength	Transmission[b]	Fractional loss per step
0.75	1.00	---
2.25	0.92	0.081
3.75	0.86	0.070
5.25	0.79	0.073
6.75	0.75	0.055
8.25	0.71	0.059
9.75	0.67	0.057
11.25	0.63	0.057
12.75	0.60	0.048
14.25	0.57	0.043
15.75	0.54	0.059
17.25	0.50	0.070
18.75	0.47	0.067
20.25	0.44	0.056

[a]Wilks Scientific 20-m Long Path Gas Cell.
[b]Transmission values relative to transmission at 0.75 m.

mirrors at one end (the in-focus end) and objective mirrors at the
other (the out-of-focus end). The optical arrangement results in
wider spacing of the images on the field mirror for which two potential advantages are claimed.

1. More images can be packed on the field mirrors before overlapping occurs than is the case with three-mirror systems.

2. Angular adjustment of the objective mirrors is less critical
while the performance at long path is more reliable. In fact, for
moderate-length cells (20 to 40 m), little or no advantage is realized since the image-packing limitation can be made to coincide approximately with the reflection loss limitation. .Simple rugged construction has made the three-mirror system by far the most reliable
long path system. The increased mechanical complexity and poorer
path-to-volume ratio of the eight-mirror system make it unattractive
for routine work. However, the optical approach taken in the design
of the eight-mirror cell may be extended to large mirror arrays,
which seem to offer some real advantages for very long path systems
and for use with laser sources.

2. Spectrometers

All types of laboratory spectrometers can be used for air contaminant analysis with the usual situation generally applying, that
greater analytical power is provided by higher-performance instruments.
However, one aspect of sampling for air analysis, namely, the need
to sample continuously in many cases, has lead to the widespread
use of small laboratory spectrometers and miniature portable spectrometers. The portability allows the spectrometer to be taken to
the sample site and used as a survey instrument providing information
that could not be obtained by a sample collection technique.

B. Minimum Detection Limits of Infrared Methods

Factors affecting the sensitivity of IR methods have already
been discussed (see Chap. 5 and Sec. A.1.a).

If we make the usual assumption that sensitivity is limited by the signal-to-noise (S/N) ratio of the analytical system, then we must agree on what level of signal is required in order that it be distinguishable from the noise. This signal level is usually set at some multiple of the peak-to-peak noise. Typical multiples are 1/6, 1/2, 1, 2, and 3; the choice within this range is rather arbitrary.* The concentration of a particular substance required to produce a signal just large enough to meet the detectability criterion chosen is known as the *minimum detectable concentration*. Table 3 shows typical minimum detection data for a variety of IR methods. The direct method is simply to measure the concentration which, in fact, produces a signal that is distinguishable from the noise. Because direct measurement often involves rather difficult sample handling, the somewhat less satisfactory method of extrapolation is commonly used. That involves preparation of a calibration curve over a range of low but readily measurable concentrations. Extrapolation of this curve to the signal level set as the detection limit gives the minimum detectable concentration.

C. Calibration Methods for Gas Analysis

The IR spectra of gases do not, in general, obey the Beer-Lambert law (see Chap. 5). It is always necessary, therefore, to construct a calibration curve which is usually a plot of absorbance vs. concentration. In environmental studies one is almost always interested in measuring minor components of a gas mixture whose concentrations range from about 1 ppb to 1 to 2% although occasionally concentrations in the 10 to 20% range are encountered, e.g., in the measurement of CO_2 from combustion processes. A variety of

---

*The choice of the minimum detectable signal level should properly be chosen on the basis of a required confidence level. Taking the minimum detectable signal level as equal to or greater than the peak-to-peak noise corresponds to a confidence level greater than 99% based on five measurements. See Ref. 3.

TABLE 3

Detection Limits (ppm) of Selected Gases for Various IR Instruments[a]

Gas	λ (μm)	Zeller and Pattacini[b]	Stewart and Erley[c]	Wilks[d]	Hanst et al.[e]
Acetone	8.2	-	5	0.1	-
Acetylene	13.6	-	-	-	0.00002
Ammonia	10.4	-	20	0.2	0.0002
2-Butanone	8.5	-	10	0.1	-
Carbon dioxide	4.25	10	5	0.05	-
Carbon monoxide	4.6	200	20	0.2	0.001
Carbon tetrachloride	12.6	-	0.5	0.06	-
Chloroform	13	-	1	0.06	-
Dichlorodifluoromethane	9.3	-	1[f]	0.08	-
Ethylene	10.55	50	5	0.7	0.0002
Formaldehyde	3.6	-	-		0.001
Hexane	3.4	5	5	0.02	-
Hydrogen chloride	3.37	-	-	0.5	0.001[g]
Hydrogen sulfide	7.7	-	-		0.05
Methane	3.3	50	-	0.2	0.0006
Methylene chloride	13.3	-	2	0.2	-
Nitrous oxide	7.68	-	25	0.8	0.002
Nitric oxide	5.26	-	-		0.0002
Nitrogen dioxide	6.2	-	-	-	-
Phosgene	11.7	-	1	-	0.0001
Sulfur dioxide	8.6	20	15	0.5	0.0002[h]
Vinyl chloride	10.63	-	10	0.7[i]	-

[a]Detection limits vary with resolution as well as pathlength, so that simple relationships based on pathlength ratios should not be expected. All measurements were made at 1 atm total pressure.

[b]M. V. Zeller and S. C. Pattacini, Infrared Bulletin No. 28, Perkin-Elmer Corp., Norwalk, Conn. Measurements were made with a Perkin-Elmer 727 spectrometer using a 1-m path.

[c]R. D. Stewart and D. S. Erley [15]. Measurements were made with a Perkin-Elmer 221 spectrometer using a 10-m path.

experimental arrangements have been proposed for making gas mixtures with accurately known concentration and transferring them to IR absorption cells. We will describe several flow and static systems which combine simplicity with good accuracy. Descriptions of other systems may be found in Refs. 4 and 5.

1. Calibration Standards

a. *Compressed Gas Mixtures*. Calibrated gas mixtures are available from commercial suppliers for a wide variety of gases in concentrations from ppm to percent. Mixtures can also be made in the laboratory if a pressure vessel and a source of compressed diluent gas are available. Mixtures may be made up gravimetrically but final concentrations should always be checked by independent analysis. Compressed gas mixtures are convenient for single-point checking of instrument calibration but become rather cumbersome and expensive when a large number of different concentrations are required. In addition, minor components of the mixture may be reduced in concentration as the mixture ages due to adsorption or reaction at the cylinder walls and that this would be undetected. Adsorption can also cause concentration to change as a function of total pressure in the cylinder. These potential problems are largely avoided by methods which involve producing calibrated mixtures from pure components at the time of use.

TABLE 3 (footnotes continued)

---

[d]Wilks Scientific Corporation, "OSHA Concentration Limits for Gases" (Wall Chart). Measurements were made with a Wilks Miniature Infrared Analyzer using a 20.25-m path.

[e]Philip L Hanst, Allen S. Lefohn, and Bruce W. Gay, Jr., *Applied Spect.*, *27*, 188 (1973). Measurements were made with a Digilab FTS System, Model 496 using a 500-m path.

[f]Analytical wavelength 10.85 μm.

[g]Analytical wavelength 3.55 μm.

[h]Analytical wavelength 7.35 μm.

[i]Analytical wavelength 10.9 μm.

b. *Static Mixtures In Plastic Bags.* A variety of polymer
films are suitable for use in sample bags. These are available either
as ready-made sample bags or as heat-sealable films from which bags
may be fabricated.* Suitable materials include polyethylene, Saran,
nylon, Teflon, and Mylar. When making calibrated mixtures the diluent
gas must be measured into the bag either by measuring flow rate and
time or by using a totalizing meter. If a wet-test totalizing meter
is used, it should be followed by a drying train to remove moisture.
In this case, the measured volume should be corrected to that of the
dry gas according to

$$V_d = V_w \frac{p - p_w}{p} \tag{4}$$

where V_d = volume of dry gas

V_w = volume of wet gas as measured by wet-test meter

p = barometric pressure

p_w = vapor pressure of water at temperature of test

The sample gas or liquid may then be introduced with an accurate
syringe. This is done either by having a septum attached to the bag
or by injecting directly through the wall of the bag and closing the
hole left by the needle with a piece of tape. The bag should be
kneaded to ensure mixing and then the mixture used promptly. Bag
methods should not be used unless preliminary experiments have shown
no significant reaction or permeation of the bag by the sample. It
is a simple matter to run spectra of samples from the bag to check
for a decrease in absorption intensity with time indicating a loss
of sample.

c. *Blending Systems.* Gas blending systems which measure the
flow of a sample stream into a diluent gas stream of known flow rate
can be used to provide a continuous source of a calibration mixture.
A simple, single-stage system using rotameters has been described by
Nelson [4]; see Fig. 4. It is claimed that such systems are capable

---

*Many commercial heat-sealing tools are available but a small
chisel-pointed electric soldering iron on a variac will work well.

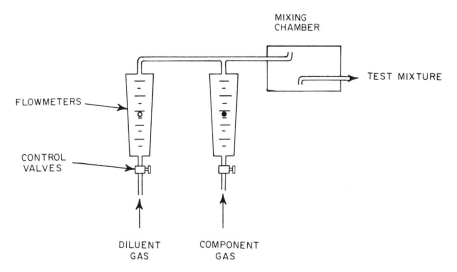

FIG. 4. Schematic diagram of blending system for single component in diluent gas.

of dilutions of 1000:1 with accuracies of about 3% [4]. When lower concentrations are required, a calibrated mixture may be used in place of the pure material or extra dilution stages may be added to the system. Complete blending systems are available commercially.

 d. Closed Loop Calibration. A fast and simple calibration system developed in our laboratory is illustrated schematically in Fig. 5. The system consists of a stainless steel bellows pump whose inlet and outlet are connected to the inlet and outlet of the gas cell. The line leading to the cell inlet includes a septum through which samples may be injected with a syringe. Usually the syringe is filled with a sample large enough to produce the highest concentration mixture required. This sample is then injected in steps to produce the individual points on the calibration curve. A typical experimental run is shown in Fig. 6, and the resulting calibration curve in Fig. 7. When gaseous samples are used with gas samples, it is important that the syringe have a very fine needle to prevent the sample's being pulled from the syringe by the flowing diluent gas. As Fig. 6 shows, the signal level is established very quickly at each

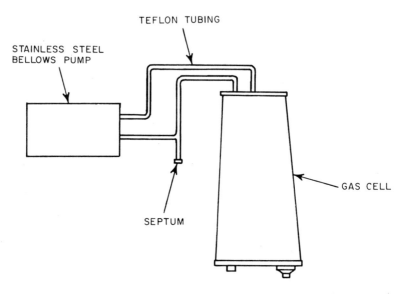

FIG. 5.   Schematic diagram of the closed loop calibration system.

point.   Although mixing within the cell is known to be considerably slower than the indicated rise time, there is no change from the original level with further mixing.   This can only mean that, provided the molecules are in the light beam, it does not matter whether they are uniformly distributed or not.   Another aspect of this phenomenon is discussed in Sec. II.D.2.   A further advantage of the closed loop system is the way in which it shows when sample loss (e.g., by adsorption, reaction, or leakage) is occurring in the system.   Figure 8(b) shows instrument read-out vs. time after an injection of SO_2 into a long path cell containing a piece of bare aluminum. Figure 8(a) shows the same experimental arrangement except that the unprotected aluminum has been removed.   Two points are clear: (1) removal of sample by adsorption or reaction is obvious in this calibration system and (2) quite good calibration information can be obtained for moderately adsorbing samples by extrapolating the absorption vs. time curves [like Fig. 8(b)] back to the time of injection. Such calibration information is, of course, only useful for analyses done under flow conditions.

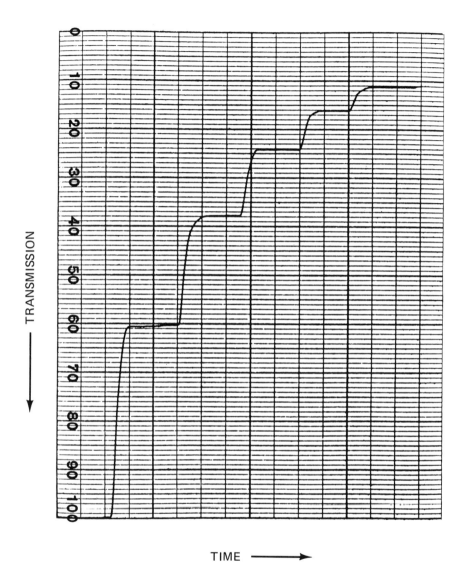

TIME ⟶

FIG. 6.   Chart record of hexane calibration experiment using the
closed loop system.   Pathlength 9.75 m.   Concentration 33 ppm per
step.   Spectral slit about 50 cm^{-1}.

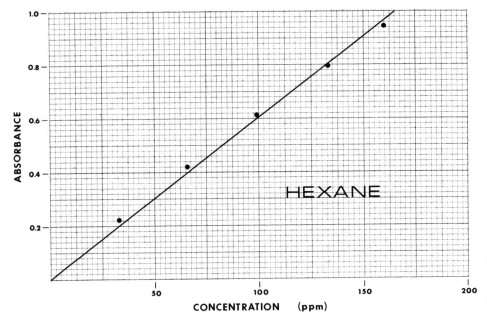

FIG. 7. Calibration curve plotted from the data of Fig. 6.

Calibration by an injection method requires that the cell volume,
including all external plumbing which contains the sample, be known.
In the case of the closed loop system, the volume external to the
cell is small and its volume may be estimated geometrically with
sufficient accuracy. Cell volume can be measured in several ways:
(1) by directly by filling the cell with water after replacing all
moisture sensitive windows with metal discs, (2) by measuring the
gas volume required to fill the evacuated cell using a gas buret or
totalizing flow meter, (3) by measuring the pressure change when the
evacuated cell is connected to a vessel of known volume of about the
same capacity as the cell (note that this method will require pressure
readings to better than 1 torr), (4) by injecting a known amount of
some volatile substance into the cell and by analyzing the resulting
mixture with an independent method. Once the cell volume is known,
concentrations may be calculated from injection sizes as follows:

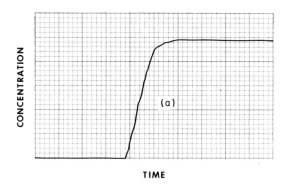

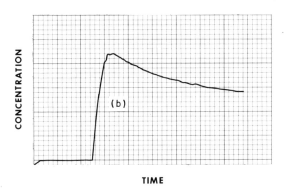

FIG. 8. Chart record showing sample loss with the calibration system. (a) Normal behavior for injections of SO_2; (b) sample loss
induced by placing a piece of unprotected aluminum within the sample
cell.

For liquids: $C(ppm) = \dfrac{\rho V_s}{M} \left[\dfrac{RT}{P}\right] \dfrac{10^3}{V_c}$ (5)

where $C(ppm)$ = gas concentration in parts per million

$\quad\quad\quad$ V_s = volume of liquid injected (μl)

$\quad\quad\quad$ ρ = liquid density in (g/cc)

$\quad\quad\quad$ M = molecular weight

$\quad\quad\quad$ $\left[\dfrac{RT}{P}\right]$ = molar volume of a gas = 24.45 at 25°C

$\quad\quad\quad$ V_c = cell volume (liters)

For gases:

$$C(ppm) = \frac{V_g}{V_c} \qquad (6)$$

where V_g = volume of gas injected (μl)

Injection sizes and resulting concentrations are shown in Table 4 for several materials.

    e. Permeation Tubes. It has been shown [6] that readily liqui-fied gases, such as SO_2, and volatile liquids, such as n-pentane, when sealed in tubes of Teflon FEP will permeate through the tube walls at a rate which depends on temperature. By placing such a tube in a diluent gas stream of known flow rate a low-concentration mix-ture may be produced. The concentration of the mixture may be ad-justed by changing the flow rate of diluent gas or by changing the temperature of the permeation tube. This method is capable of pro-ducing very accurate calibration mixtures. Certified permeation tubes for SO_2 are now available from the National Bureau of Standards and tubes for many other materials are available from commercial suppliers. Good discussions of the method can be found in Refs. 4 and 5.

    Permeation methods have not been much used in IR gas analysis because permeation rates near room temperature are such that ppm range mixtures are produced at diluent gas flow rates in the ml/min range. Reasonably efficient sample exchange in long path gas cells requires flow rates in the 10- to 50-liter/min range (see Sec. C.3.c).

    2. Construction and Use of Calibration Curves

    As has been pointed out in Chap. 5 calibration curves for IR methods are almost always nonlinear. This is particularly true in the gas phase and arises because the individual rotational lines of a vibration-rotation absorption band have widths which are narrower than the band pass of the spectrometer. The exact nature of this effect has been discussed in detail elsewhere; see, for example, Nielsen et al. [7] and Jamieson et al. [8]. From an analytical point

TABLE 4

Concentrations of Calibration Mixtures Produced

in a 5.6-liter Cell by Injection

Sample	Injection size (μl)	Concentration (ppm)
Acetone (liquid)	1	60
Acetone (sat. vap. at 22.7°C)	100	4.7
Carbon monoxide (gas)	100	17.8
Hexane (liquid)	1	33
Toluene (liquid)	1	41
Toluene (sat. vap. at 18.4°C)	100	0.47
Vinyl Chloride (gas)	100	17.8

of view, it is important only to catalog the nature and range of the effects that this situation can have on the intensity of the observed spectrum. We note the following:

1. Calibration curves (absorbance vs. concentration plots) for gases at low concentration can vary from almost linear to strongly curved but will always be concave downward. Both the amount of curvature and the average slope of the calibration curve may vary markedly with spectral resolution.

2. Any change in the line shape of the rotational lines will affect the observed intensity. At atmospheric pressure the rotational line shape depends on the nature and frequency of molecular collisions. The apparent spectral intensity will, consequently, be affected by changes in temperature and pressure (which affect collision frequency) and by the nature of the diluent gas (which affects the nature of molecular interactions).

3. In general, the most strongly curved calibration curve will show the greatest sensitivity to temperature, pressure, and nature of diluent gas. The consequences of these facts which affect calibration and sampling will be considered in the following discussion.

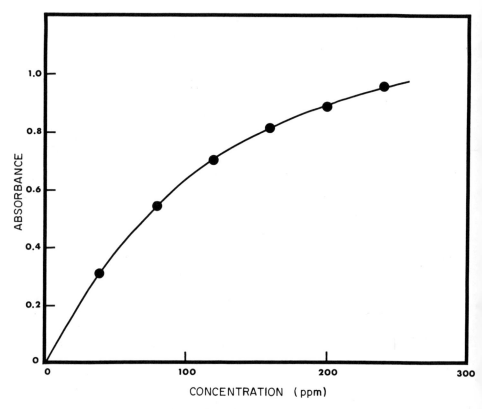

FIG. 9. A versus C calibration curve for tetrachloroethylene.
Pathlength = 5.25 m; λ = 10.9 μm.

  a. *A versus C Plots:* Spectroscopic calibration data are usually presented as a graph of absorbance vs. concentration, such as that
shown in Fig. 9 for tetrachloroethylene. Absorbance is the spectroscopic quantity most simply related to concentration and gives plots
with the least curvature. However, in certain situations other plots
are useful.

  b. *A versus CL Plots:* Figures 10 and 11 are plots of absorbance
vs. the product of concentration and pathlength. As the plots clearly
show these curves are independent of the analytical pathlength over
the range of concentrations and absorbances considered; absorbance
does not depend on volume concentration but only on the total number

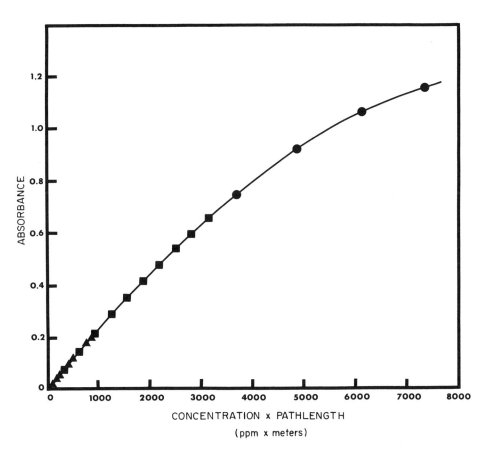

FIG. 10. A versus CL calibration curve for acetone: ● = 20.25 m; ■ = 5.25 m; ▲ = 0.75 m.

of molecules in the beam. This result is reasonable since, at these concentrations, each analyzed molecule will be surrounded by diluent molecules and will not be affected by the relatively minor changes in the numbers of these molecules in the mixture. Practically, this finding means that calibration curves determined at one pathlength of a variable path gas cell can be used at any other pathlength simply by transforming the concentration scale according to

$$C_2 = \frac{C_1 L_1}{L_2} \tag{7}$$

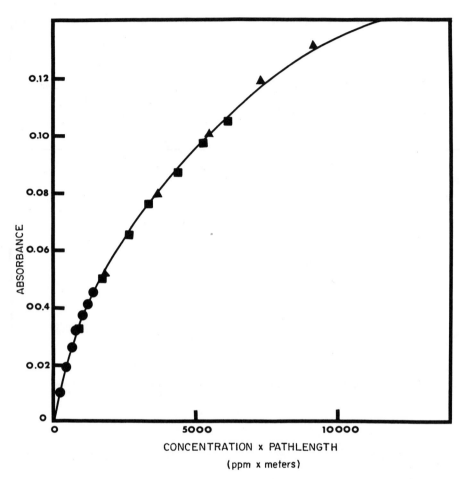

FIG. 11. A versus CL calibration curve for carbon monoxide: ▲ =
20.25 m; ■ = 9.75 m; ● = 2.25 m.

where C_1 is the original concentration for which the absorbance was
        measured at pathlength L_1.

   C_2 is the derived concentration which will give the same ab-
        sorbance when measured at pathlength L_2.

Table 5 shows typical measured and calculated data. Two points are
worth noting about the data in the table; (1) variable pathlength
gives the method wide dynamic range, (2) the transferability of calibration data from pathlength to pathlength frequently allows one to

TABLE 5

Calibration Data for n-Propyl Acetate Measured at 0.75 m and
Calculated for Other Pathlengths[a]

Concentration (ppm)	Effective concentration at 6.75 m (ppm)	Effective concentration at 20.25 m (ppm)	Absorbance
38.7	4.3	1.4	0.036
77.4	8.6	2.9	0.072
116.	12.9	4.3	0.106
115.	17.2	5.7	0.139
194.	21.5	7.2	0.171
232.	25.8	8.6	0.203

[a]Data were measured with a Wilks Scientific Corporation Miniature
Infrared Analyzer.

avoid making up calibration mixtures at inconveniently high or low
concentrations. For example, the measured data in Table 5 were obtained by making 1-μl injections into a 5.5-liter cell. Had the data
for 20.25 m been measured directly the injections would have to have
been about 0.04 μl each, a much more difficult volumetric measurement.
The use of the product cl as a single variable called the optical
thickness or optical pathlength is common practice in gas phase intensity work (see, for example, Ref. 9). The usefulness of A versus
cl plots in analytical work has been noted by Ritchie and Kulawic
[13] in a study of the natural abundance concentrations of deutero-
methane in natural gas. These authors also note the effect of different diluent gases on calibration curves, a point which is discussed further in Sec. 3.

    c. (1 - T) versus C Plots: For weak bands the quantity (1 - T)
is approximately proportional to absorbance since

    $A = -\log T$

or

    $T = \exp (-2.303 A)$

Expanding the exponential gives

$$T = 1 - 2.303A + \frac{(2.303A)^2}{2} - \ldots$$

which for small A gives

$$T \approx 1 - 2.303A$$

Figure 12 shows (1 - T) versus C and 2.303A versus C plots of the same data. The advantage of this presentation is simply that (1 - T) can be read directly from the linear transmittance readout of the spectrometer and is useful mainly when using scale expansion. It is important to remember that the base line must be set to the same value (preferrably 100%) for both calibration and analysis if values are to be read directly from the chart paper.

3. *Effect of Experimental Conditions on Calibration and Analysis*

In general, it is desirable to set the conditions identically for calibration and sample analysis. However, this is not always convenient or even possible and it is of interest to assess the magnitude of the effect that various changes in conditions will have on the analysis. We will consider separately the effects of pressure, temperature, flow rate, and nature of diluent gas. In each of these cases, we will consider first the effects of normal climatic variations and then the effect of the larger variations which might be encountered in various sampling situations.

a. *Pressure:* When a gas mixture such as contaminated air is sampled, the total number of contaminant molecules in the sample cell (and, therefore, the measured absorbance) will depend on the total pressure in accordance with the gas law. The gas law correction to be applied to measurements made at temperatures and pressures different from those used in calibration is given in Eq. (9). There is, however, an additional effect on the absorbing properties of the molecules themselves. The rotational lines of a gas phase absorption band are Lorentzian in shape except at very low pressures; that is, at normal pressures, the absorption index k varies with wavenumber according to

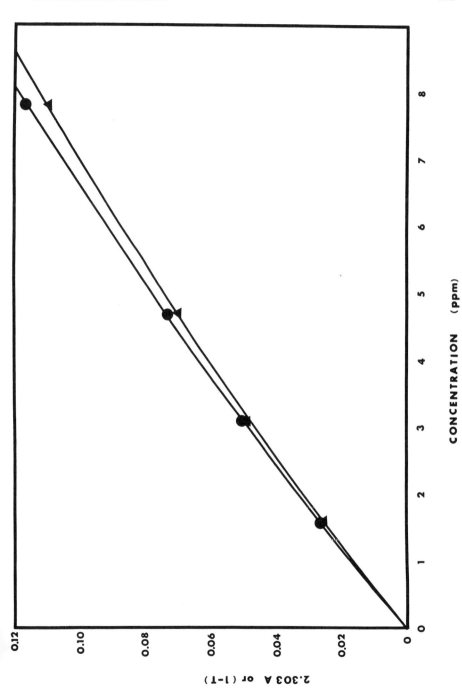

FIG. 12. Comparison of (1 - T) vs C (▲) and 2.303 A vs C (●) calibration curves for dimethylformamide.

$$k(\tilde{\nu}) = \frac{S}{\pi} \frac{\chi}{(\nu - \nu_o)^2 + \chi^2}$$ (8)

where

    $\tilde{\nu}$ is the wave number

    $\tilde{\nu}_o$ is the wave number of the line center

    χ is the half-width at half-height

    S   is the integrated intensity of the line

    $k(\tilde{\nu})$ is the absorption index at wave number

The half-width χ is proportional to pressure causing the line shape
to change as shown in Fig. 13. For an "infinite resolution" spectrometer, such as a tunable diode laser, an increase in total pressure
without changing the amount of absorber in the sample cell (e.g., by
pressurizing with a transparent gas such as N_2) would cause a decrease
in measured absorption. For conventional medium- and low-resolution
spectrometers, the effect is just the reverse as is shown in Fig. 14.
This behavior can be explained qualitatively as follows. At low pressure, the rotational lines are much narrower than the band of frequencies passed by the spectrometer. Such very sharp lines can cause
total absorption of the light for wavelengths very near the line
centers even when the total absorption within the spectrometer band
pass is quite low. Molecules toward the end of the optical path
which would absorb light of wavelengths near the line centers cannot,
since these wavelengths have been absorbed completely by molecules
preceding them. When the pressure is increased, the intensity is
decreased at the line center and increased in the wings. This causes
more of the frequencies within the spectrometer band pass to be absorbed and, consequently, causes the spectrometer to indicate a
greater absorbance. The effect is most marked for light molecules
such as CO, HCl, or CH_4 which have sharp, widely spaced rotational
lines or for polyatomic molecules with series of sharp, prominent
Q-branches, such as the methyl halides. It is clearly dependent on
spectrometer band pass and, in fact, shows dramatic differences with

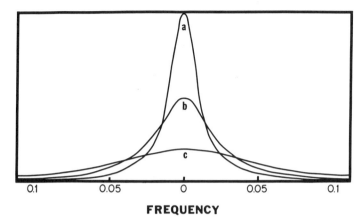

FREQUENCY

FIG. 13. Lorentzian line shape. R(4) of CO at (a) 150 torr; (b) 300 torr; (c) 750 torr.

changes in resolution. Detailed discussions of pressure-broadening effects can be found elsewhere; see, for example, Hanst [2], Nielsen et al. [7], Jamieson et al [8].

Normal climatic pressure changes at any one site can vary from about 1 to 6%. In this range, pressure-broadening effects can be ignored and the simple gas law correction can be applied. Altitude changes which can easily be 5,000 ft when going from sea level to mountain locations will cause pressure changes up to 15%. Pressure-broadening effects could then be significant and as a result recalibration would be necessary. As already noted, materials with calibration curves showing large deviations from linearity will be most sensitive to pressure-broadening effects.

From the foregoing, it is clear that spectra taken at low pressure, as is common in laboratory practice, cannot be used to calibrate methods for measurement at atmospheric pressure. Similarly, if samples are to be compressed to increase analytical sensitivity, calibration curves cannot simply be scaled by the pressure ratio but must be redetermined at the higher pressure.

Finally, reducing sample pressure can be very useful when high-resolution instruments are available. Pressure reduction will invariably cause a reduction in absorption but the accompanying

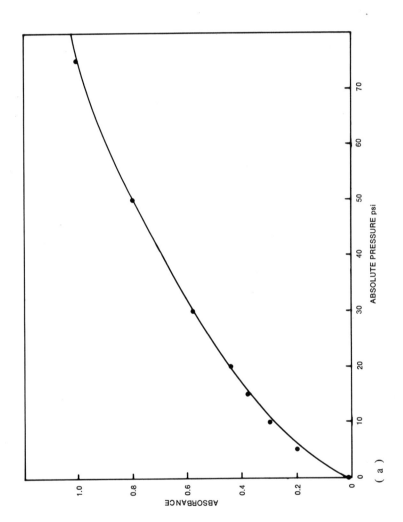

(a)

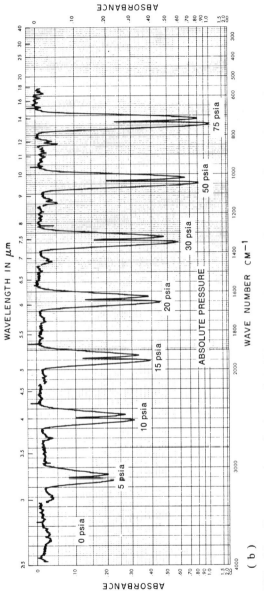

FIG. 14.   Pressure broadening of 1,000 ppm carbon monoxide in helium.   (Reproduced by permission of Beckman Instruments, Inc., Fullerton, California.)

sharpening of the lines can cause a significant increase in selectivity due to the elimination of overlap between lines from different materials. For example, Hanst and co-workers [2,10] have demonstrated that analyses are possible for materials absorbing in the 5- to 7-μm region of intense atmospheric water absorption by "looking between" the water lines.

b. *Temperature:* The amount of material in a gas cell will vary with temperature according to the gas law. Measurements made at temperatures and pressures different from those at which the calibration was made should be corrected according to

$$C_{corr} = C_{uncorr} \frac{P_c T}{PT_c} \tag{9}$$

where

C_{corr} = corrected concentration (ppm)
C_{uncorr} = uncorrected concentration (ppm)
T_c, P_c = temperature °(K) and pressure (atm) at which calibration was made
T, P = temperature °(K) and pressure (atm) at which measurement was made

Temperature effects are usually less significant than pressure effects, however, since gas expansion depends on the absolute temperature; consequently, the percentage change due to normal temperature fluctuations is small. Rotational line widths are temperature-dependent, giving rise to the same effects as line width changes due to pressure variation but to a lesser degree since line widths increase as the square root of the temperature ratio (see Ref. 7). There is an additional effect on the spectrum itself due to changes in temperature causing a redistribution of the population of the rotational energy levels. Increasing temperature changes the relative intensities of rotational lines so that the overall band contour is broadened. The effect, which is only significant at extreme temperatures, has been discussed in detail by Herzberg [11].

TABLE 6

Dynamic Pressure Drop for Air Flowing in Smooth Tubes

Tubing diameter (in.)		$\Delta P/L$	
Nom.	ID	@ 10 liter/min (atm/ft)	@ 30 liter/min (atm/ft)
1/4	0.2	0.001	0.005
3/8	0.31	0.0001	0.0007
1/2	0.38	0.00003	0.0002

    c. Flow Rate: The IR spectrum of a gas is not detectably flow rate-dependent over the range of flow rate and spectral resolution used in the gas analysis. There are, however, two secondary flow rate-dependent effects which can markedly affect the analysis of a flowing sample. They are: flow rate-induced pressure changes and sample exchange rates. Table 6 shows approximate pressure drops encountered when air at room temperature flows through straight smooth tubes. From the table we see that a sampling pump on the outlet of a sample cell used to pull a sample through both a 100-ft sample line and the cell at a flow rate at 30 liter/min will reduce the sample pressure by 50% with 1/4-in. diameter tubing and by 2% with 1/2-in. diameter tubing. Similarly, 4 ft of 1/4-in. diameter tubing on the outlet of a cell to which calibration gas is being supplied at 30 liter/min will raise the cell pressure by 2%. Clearly, care should be taken to see that sample flow into and out of the sample cell is unimpeded.

    If a sample is flowed into a sample cell, the way in which the concentration in the cell approaches that of the sample feed is given by the simple first order rate law (assuming complete continuous mixing):

$$dx = -\frac{F}{V}(x - x_f)\ dt \tag{10}$$

where

    x = concentration in the cell at time t

x_f = concentration of the sample feed

F = flow rate

V = cell volume

t = time

Solving

$$\frac{x_f - x}{x_f} = \exp\ (-Ft/V) \tag{11}$$

For the special case of no sample in the cell initially, i.e., x_o = 0.
Figure 15 shows a plot of the time required for the concentration in
a 5.5-liter cell to reach 90% of that of the sample feed vs. flow
rate. The curve is calculated from Eq. (8); the plotted points are
measured values. Note that this model predicts that the same volume
of sample will be required regardless of flow rate. Expressed in
cell volumes of sample, the requirements are 2.3 cell volumes to
reach 90%, 3 cell volumes to reach 95%, and 4.6 cell volumes to reach
99%.

D. Sample Handling

1. Samples Containing Water

a. Dew Points Above Ambient Temperature. Samples with dew
points higher than the sample cell temperature will produce liquid
water if allowed to cool in the cell. This must be avoided since
liquid water will interfere seriously with the gas measurement and
may cause severe damage to the cell windows. Condensation may be
prevented by heating the cell or by drying the sample. Cell heating
is useful when moderate amounts of water are involved; for example,
when analyzing exhaled breath to measure exposure to toxic substances
[12]. When samples contain larger amounts of water, sample drying
becomes necessary. We will consider several aspects of the sample
drying problem.

Probably the best solution to the problem of removing water
from a sample without removing the materials of interest is that
provided by the selective membrane dryer. This device consists of

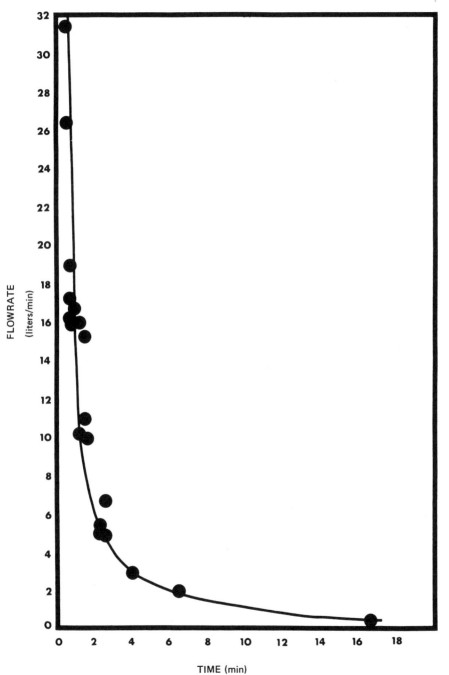

FIG. 15. Response time of 5.5-liter gas cell.

polymer tubes which are selectively permeable to water. The sample
is passed through the tubes and the water removed from it by permeation through the tube walls. Dry purge gas passing over the outside of the tubes carries the water vapor away. A commercial dryer
of this kind is available and has been described in detail in Ref.
14. The membrane, though highly selective, is not completely so and
is also permeated by other materials such as water-soluble alcohols
and ethers.

Refrigeration drying systems are used extensively to dry samples,
particularly, for combustion stack analysis. These systems may be
simple cold traps or mechanical refrigeration units but in either
case, drying is effected by simply chilling the sample to cause water
condensation. This will invariably cause sample loss but it is frequently not severe as the following analysis shows.

Figure 16 shows a schematic refrigeration dryer and gives the
composition of the inlet and outlet streams. Consider an inlet
stream consisting of a component of interest (A), water, and diluent
gas. Define

a_i = concentration of A in the feed (ml/liter)

x_A = mole fraction of A in the condensate

P_A = partial pressure of A in the product stream

α = volume of gaseous A dissolved in the condensate per liter
    of feed (ml/liter)

β = volume of gaseous water condensed in the dryer per liter
    of feed (ml/liter)

All gas volumes are to be expressed at some standard conditions, say,
25°C, 1 atm. Within the dryer Henry's law will apply, i.e.,

$$P_A = Kx_A \tag{12}$$

where

K = Henry's law constant

Combining the product and condensate stream compositions with Henry's
law gives

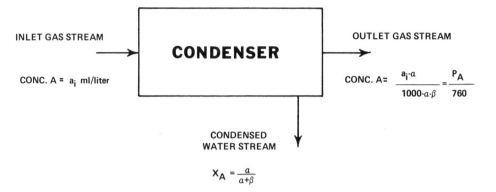

FIG. 16. Schematic diagram of refrigeration dryer showing inlet and outlet composition.

$$\frac{a_i - \alpha}{1000 - \alpha - \beta} = \frac{K}{760} \frac{\alpha}{\alpha + \beta} \tag{13}$$

Solving with suitable approximations gives the fractional loss of component A:

$$\frac{\alpha}{a_i} = \frac{760\beta}{1000K - K\beta + 760\beta} \tag{14}$$

Table 7 gives values of β for various sample dew points and condenser temperatures. Table 8 gives the loss of various materials when a sample with a dew point of 50°C is cooled to 5°C.

Water condensation can also result from compression of samples but this is experimentally more difficult. It also causes greater sample loss since solubility is proportional to pressure (according to Henry's law) but increases relatively slowly with decreasing temperature. The preceeding analysis can be used to estimate sample losses in compression drying as well as refrigeration drying, but the value of β must be evaluated from vapor pressure data rather than taken from Table 7.

    b. Dew Points Below Ambient Temperature. Samples with dew points below ambient temperature do not suffer from the difficulties of condensing water, but spectral interference from water vapor may

TABLE 7

Volume of Water Vapor (β) Condensed When a Sample Stream is Cooled[a]

Drier temperature (°C)	Sample dew point			
	30°C	40°C	50°C	60°C
0	31.6	67.2	116	192
5	33.6	64.7	114	190
10	30.1	61.4	111	187
15	25.4	56.9	107	183
20	19.6	51.9	103	181

[a]All volumes in milliliters at 25°C, 1 atm.

TABLE 8

Sample Loss in Refrigeration Dryers[a]

Compound	Henry's law constant	
	K(5°C)	% Loss
CO	30.02×10^6	0.0003
CH_4	19.69×10^6	0.0005
NO	14.63×10^6	0.0007
C_2H_4	4.96×10^6	0.002
N_2O	0.902×10^6	0.01
COS	0.891×10^6	0.01
CO_2	0.666×10^6	0.01
C_2H_2	0.63×10^6	0.02
SO_2	0.015×10^6	0.6
NH_3	1.562×10^3	6
HCl	1.0×10^3	∿9

[a]Sample dew point 50°C, final temperature 5°C.

be a problem. Water vapor absorbs significantly in the regions 4000 to 3500 cm^{-1} (2.5 to 2.9 μm), 2000 to 1350 cm^{-1} (5 to 7.4 μm), beyond 650 cm^{-1} (∿15 μm), and to a lesser extent in most other regions of the spectrum. When instruments with good resolution are used, the water vapor absorption may be compensated optically as

shown in Fig. 18 (Section II.E.1) or by the computer ratioing methods
described by Hanst [10]. For sample concentrations in the ppm or
higher ranges, water vapor interference is not significant except in
the regions of high absorption. For measurements in these regions
the selective membrane drier [14] may dry samples sufficiently to
allow uncompensated measurements and will certainly make compensation
easier.

For measurements in the sub ppm range, water vapor interference
will be significant in most regions of the spectrum. Drying of such
samples is not usually possible without significant sample loss.
Optical or computer compensation methods work well but one other
method is worth considering. Most organic vapors are efficiently
removed from air by activated carbon filters, such as those used on
canister-type gas masks, while water and carbon dioxide are not. If
such a filter is used to remove the component of interest from part
of a sample, the water and carbon dioxide background can be determined. Admitting the sample without filtration then allows the required measurement to be made by difference.

c. *Drying Samples for Pressurization.* The intensity of the
spectrum from a gas sample can be increased by compressing the sample
in the gas cell. As the previous discussion points out, the intensity increase will involve both the concentration increase and the
effect of pressure on the spectral lines. A difficulty with using
elevated pressure to increase the sensitivity of ambient air measurements is that raising the pressure of the sample quickly brings it
to saturation with respect to water. Further pressurization would
cause water to condense. For example, an air sample with a relative
humidity of 50% would be saturated if compressed to 2 atm and further
compression would cause condensation. As we have already discussed,
a compression of this sort could be used as a sample drying technique,
except that it tends to cause excessive sample loss. In practice,
drying before compression is a more useful procedure. Baker [16]
has shown that the selective membrane dryer is suitable for this
purpose. Samples containing a wide variety of organic vapors may be

dried selectively so that compression to 10 atm does not cause condensation. Baker has also shown that a second dryer on the high-
pressure side of the compressor can be operated with pressurized
purge gas to dry samples sufficiently to eliminate water interference even in the regions of the spectrum where water absorbs strongly.

2. *Sampling Methods*

A large number of gas-sampling methods have been described in
Ref. 5. However, most of these methods are unsuited to sampling for
long path gas cell analyses since they provide insufficient sample
volume. If samples are concentrated, a small vessel can be used and
the dilution involved in admitting the sample to a large cell taken
into account. However, for most ambient air work large dilutions
are not tolerable. The most generally useful sampling devices are
the sample bag and the long path cell itself.

a. Sample Bags: Plastic bags have already been described as
mixing vessels for calibration mixtures in Sec. C.1. The same considerations of chemical inertness apply for samples as for calibration mixtures. Any bag method should be carefully checked for sample
disappearance with time before an attempt is made to obtain analytical results. In addition to reaction with the sample bag, the possibility of reaction between components of the sample must be considered.
Bag sampling methods are particularly useful when the qualitative
power of the laboratory spectrometer is required. For preliminary
analyses, bags provide a simple way of separating gases from solid
and liquid particulate matter without elaborate sample conditioning.
An example of a simple qualitative comparison of auto exhaust using
bag samples is shown in Fig. 17.

The chemical inertness and impermeability of tubing used in
sampling are as important as the properties of the bag itself. Most
ordinary laboratory tubing of rubber or plasticized polyvinylchloride
is entirely unsuitable for sampling organic vapors. Saran and nylon
tubing work well for sampling many materials, although Saran absorbs
nitrobenzene readily. Polyethylene and polypropylene are also useful
in special cases but should always be tested. Teflon is the most

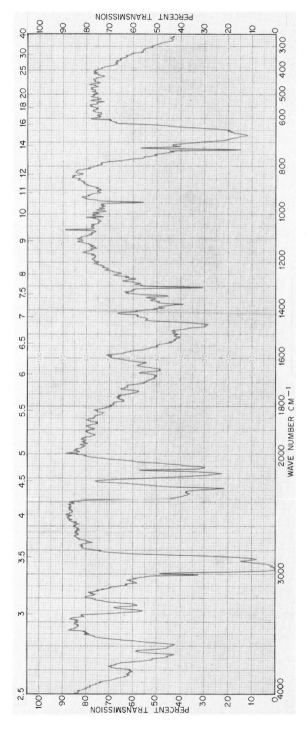

FIG. 17(a). Infrared spectrum of automobile exhaust from a 1967 Volkswagen. Reference cell, 20.25 m; sample cell, 20.25 m; gas phase; 1 atm; diluted with ambient air. Compare this spectrum with those of Figs. 17(b) and (c).

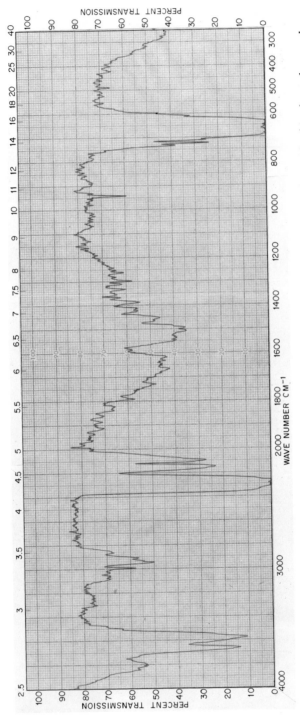

FIG. 17(b).   Infrared spectrum of automobile exhaust from a 1973 Plymouth Duster (27,000 miles), taken under the same conditions as Fig. 17(a).

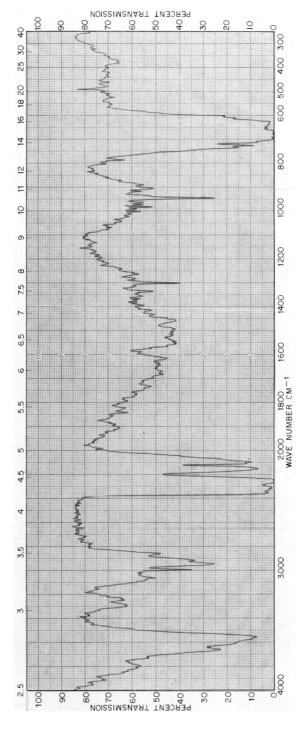

FIG. 17(c).   Infrared spectrum of automobile exhaust from a 1974 Honda Civic (350 miles), taken under the same conditions as Fig. 17(a).

universally suitable sample line material from a chemical point of view. The closed loop calibration system already described provides a convenient way of checking tubing materials with specific samples. By charging the cell with a sample and closing the loop with a substantial length of the tubing to be tested, the sample can be circulated through the tubing and the spectrum measured continuously to detect any decay in the sample concentration. One of the major difficulties in bag sampling, or any sample-collecting methods, is in knowing where to collect the sample.

    b. Continuous Sampling Methods: The introduction of small portable spectrometers has allowed on-site analysis to be used for many air contaminants. On-site measurements not only give the analyst immediate results, a great aid in sampling, but frequently provide important information which is unobtainable by sample collection methods. For example, toxic vapors in a chemical process building are rarely found to be uniformly distributed. Rather, they form high-concentration streams originating at the vapor sources within the building and which are controlled by subtle air currents. An air current which is barely discernable flowing past an open vessel containing a volatile solvent can keep the area a few inches upstream of the vessel solvent vapor-free — less than 1 ppm — while concentrations of hundreds of ppm may be found downstream.

Continuous analysis also allows the study of time-dependent phenomena. For example, the release of vapors into the air at certain stages in a process is usually very obvious when a continuous record is made and may be almost impossible to detect from the results of collected samples.

The combination of a continuous quantitative analysis method with a laboratory qualitative method is particularly powerful in tracking down elusive contaminants. A common problem in qualitative analysis is finding a sample which is concentrated enough to provide a conclusive analysis. A continuous method allows the analyst to search out a concentrated sample which can then be returned to the laboratory for qualitative identification.

E. Analysis of Multicomponent Mixtures

Several features of air contaminant analyses combine to make the analysis of multicomponent mixtures a much less formidable task than might be supposed:

1. Since the mixtures consist of low concentrations of contaminants in air, only the strongest bands of the contaminant spectra appear with significant intensity. This simplifies the spectra and greatly reduces the chance of mutual interference.

2. Gases are generally very transparent except in the immediate region of absorption bands, so that there is no significant interference problem unless bands actually overlap.

1. Atmospheric Absorptions

The most common interference problem encountered in atmospheric sampling is that of atmospheric moisture and CO_2 interfering with analyses for contaminants. Figure 18 shows the spectrum of atmospheric water and CO_2 at 20.25 m pathlength. The upper trace shows the optical compensation possible when two identical long path cells are used. With a medium- or high-resolution spectrometer sufficient energy is available between the sharp intense water absorption lines to allow spectra to be recorded in these regions. This is not always the case for low-resolution spectrometers, and measurements in this region may require sample drying. Fortunately, most organic compounds have characteristic absorptions in the 1200- to 650-cm^{-1} region where water absorption does not interfere significantly with analysis down to the ppm level. For analyses in the sub ppm range water vapor absorption is significant in most regions of the spectrum. Interference may be removed by optical compensation as discussed above, by the computer methods described by Hanst [10], or by using a difference method which subtracts out the water contribution. The many strong lines in the water spectrum provide a sensitive test of water compensation making this an excellent method when the necessary equipment is available. When compensation is not possible, organic vapors can conveniently be analyzed by using an activated carbon filter, such

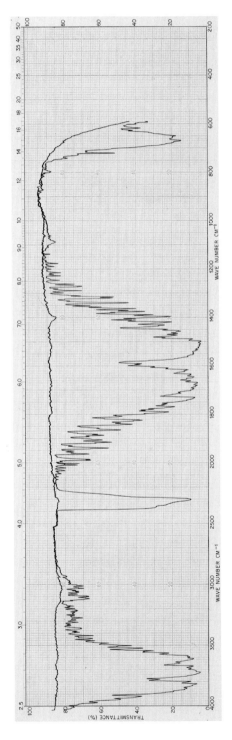

FIG. 18. Atmospheric absorption at 20.25-m path and optical compensation using two identical cells.

as a gas mask cannister, to remove just the organic material giving an accurate measure of the humid air background. It is important to humidify the filter first, since dry activated carbon does retain some water but once humidified it no longer affects the water content of the sample. Moreover, humidification increases the retention of organic vapors by the filter. Analyses of the insecticide DDVP have been performed down to the ppb level by this method [17].

   2. Multicomponent Analysis Without Interference

   As already noted, a large number of multicomponent analyses are possible simply by measuring the intensities of analytical bands for the individual components. Figure 19 demonstrates an analysis of this type showing that the absorption bands, though quite close in frequency, do not affect each other.

   3. Multicomponent Analysis With Interference

   Occasionally a multicomponent analysis is encountered in which at least one of the components cannot be determined independently. A mixture containing trichloroethylene and tetrachloroethylene is such a case; trichloroethylene interferes with the tetrachloroethylene analysis. Figures 9 and 20 give the necessary information for this analysis. Measurement of the absorbance of the band at 11.8 µm from Fig. 20 gives the concentration of trichloroethylene and its contribution to the absorbance at 10.9 µm. The measured absorbance at 10.9 µm less the trichloroethylene contribution gives the absorbance due to tetrachloroethylene, which can be used with Fig. 9 to give concentration. When there is mutual interference, corrections can usually be estimated adequately by ignoring the interference initially.

F. Remote Sensing Methods

   Remote sensing systems for measurement of general air pollution and for measurement of pollution sources from distant points have been under development for some time. Recent programs including IR laser systems, remote Raman, conventional heated source IR systems, and solar source systems have been reviewed by Harney et al. [18]. Some typical programs are described briefly in the following sections.

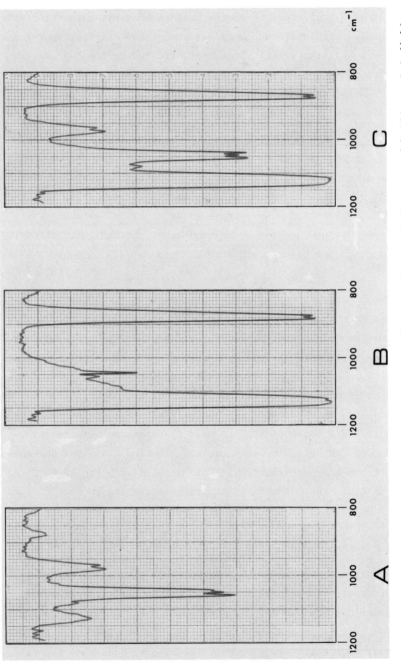

FIG. 19.   Independence of analytical bands of 1,1-dichloroethane and dioxane.   (a) 104 ppm 1,1-dichloroethane; (b) 104 ppm dioxane; (c) 104 ppm 1,1-dichloroethane plus 100 ppm dioxane.   Pathlengths 20.25 m, spectral slit ~3 cm^{-1}.   Analytical bands: 1,1-dichloroethane, 980 cm^{-1}; dioxane, 883 cm^{-1}.

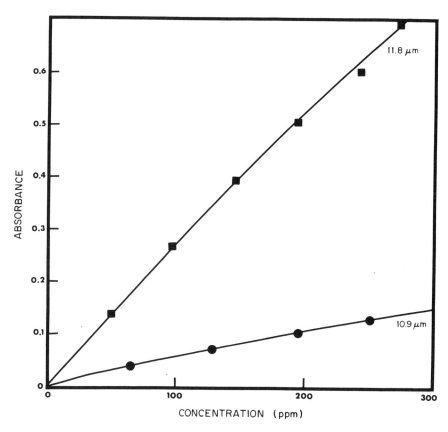

FIG. 20.   Calibration curves for trichloroethylene at 10.9 and 11.8 μm, 5.25-m path, spectral slit ~20 cm^{-1}.

1.   Remote Raman

A remote pulsed laser Raman system has been in development for several years [19]. The system includes a pulsed ruby laser, a receiving telescope to collect back-scattered Raman radiation, and a polychromator for spectrum analysis. Advantages of a Raman system for remote measurement include relative freedom from atmospheric interference, internal standardization for quantitative analysis (using atmospheric nitrogen), ability to measure almost all pollutants and, by using time gating, the ability to define the region of space being sampled. The principal disadvantage of the technique is the inherent weakness of the effect. Detection limits are claimed to be in the ppm range while comparable IR systems will detect to the ppb range.

2. *Laser Infrared Systems*

Both fixed wavelength and tunable IR lasers are being used to monitor atmospheric pollutants by absorption. Hanst [2] has cataloged spectra of a number of air pollutants showing coincidences between pollutant absorption lines and emission lines of various gas lasers. Sufficient coincidences occur that most common pollutants could be detected with an appropriate gas laser.

Snowman and Gillmeister [20] have reported the field testing of a long-range laser system for pollution measurements. The system consists of a CO_2 laser and IR detector mounted close together. The laser is aimed at a retroreflector 1 or 2 miles away which returns the beam to the detector. Various pollutants along this path can be detected by careful comparison of the absorption of several different lines from the laser. A serious problem with this kind of measurement is the effect of refractive index variations of the atmosphere due to temperature variations along the path. These random fluctuations deflect the beam and cause noise or "scintillation" in the signal. Snowman and Gillmeister found that scintillation was most severe at midday and midnight under clear skies. Scintillation was normally least at dusk and was reduced by wind and overcast skies. Detection limits over a 2-mile path during conditions of maximum scintillation were in the 10- to 20-ppb range.

Hinkley [21] has described a tunable diode laser system with which extremely high-resolution spectra may be measured. This system is being used for long-range absorption measurements of pollutants. Sensitivity has been estimated at 0.3 ppm/m of path for strong absorbers such as ammonia and sulfur hexafluoride.

3. *Solar Source Systems*

IR spectroscopy using the sun as a source for airborne spectrometers has been used to measure pollutants in the upper atmosphere.

Murcray et al. [31] have used a balloon-borne grating spectrometer
to detect the presence of nitric acid in the upper atmosphere.
Schindler [32] has described a Fourier transform spectrometer suitable for airborne operation which should improve measurements of this
kind.

III. WATER ANALYSIS

Direct measurement of water pollutants by vibrational spectroscopy
is rarely possible. Water is too strong an absorber of IR to allow
contaminants at the concentrations of interest in pollution studies
to be detected. In principle, Raman measurements directly on water
samples should be possible but, in practice, fluorescence problems
often make the detection of trace components difficult or impossible.
Practical methods usually involve a separation of the contaminant
from the water sample followed by spectral analysis. For qualitative
analysis, the IR and Raman spectra will usually be independently useful. For quantitative analysis, once a separation has been accomplished, the IR is usually more easily applied and is less expensive.
That will not necessarily be the case for materials which are weak
IR absorbers and strong Raman scatterers, however. In the following
sections, several sample-handling techniques are discussed which
could be used to prepare samples for either IR or Raman measurements
although they have more commonly been used in the IR.

A. Dissolved Organic Contaminants

A number of organic materials in water can be analyzed either
by direct extraction into carbon tetrachloride or Freon or by conversion to a readily extractable species prior to extraction. The
extraction both separates the organic material from the water and
provides up to 100-fold concentration. Detailed methods are available for phenols and petroleum-based oils and these are discussed
below. However, the method should be applicable to other materials,
such as vegetable oils, pesticides, organic solvents, etc.

Sewage treatment plants commonly determine the hexane-extractable fraction of both their inlet and outlet streams. This procedure normally involves extraction with hexane, evaporation to dryness, and weighing of the residue. It would seem that a good deal of useful information might be obtained from the IR spectrum of either the hexane extract or the residue that remains when the extract is evaporated on an ATR plate. To the author's knowledge, no studies of the kind have been reported.

1. Oil in Water

A standard method is available [22] for the analysis of oil in water. The method involves extraction with CCl_4. The amount of organic material is determined by measuring the absorbance at 3.4 μm using either a 1-cm or 5-cm quartz near-IR cell. Typical detection limits are given in Table 9. Calibration is best done by using an oil sample from the source of the water contamination. When this is not available, a synthetic sample consisting of 37.5% cetane, 37.5% iso-octane, and 25% benzene is commonly used. The absorbance measurement at 3.4 μm tends to underestimate aromatics but this does not usually introduce a significant error, since the aromatic contribution in the calibration sample is similarily underestimated.

A useful variation of the standard method [23] has been developed in which screwcap sample bottles with foil-lined caps are used to collect samples and the extraction is performed in the sample bottle. The carbon tetrachloride layer is withdrawn using a pipette with a filter paper covering on the tip. This method eliminates undesirable sample transfers and substantially reduces the cost of the equipment required.

Because such small quantities of hydrocarbon are being detected, it is essential that no hydrocarbon grease or hydrocarbon polymers (such as polyethylene or polypropylene) be used in any of the sample-handling equipment. It is equally important that high-quality carbon tetrachloride be used and that solvent from the same source be used for both calibration and unknown samples. Glassware must be cleaned carefully to avoid contamination; a detergent washing method has been tested and found to be satisfactory for this purpose [24].

TABLE 9

Detection Limits for the IR Determination of Oil in Water[a]

Sample volume (ml)	CCl$_4$ volume (ml)	Detection limit (ppm)
1000	20	0.2
1000	60	0.6
1000	100	1
2000	20	0.1
2000	60	0.3
2000	100	0.5
3000	20	0.1
3000	60	0.2
3000	100	0.3

[a]Measured with Wilks Scientific Miran-I Analyzer, cell path 5 cm.

2. Phenols in Water

A method for the determination of phenols in water has been developed [25] based on the fact that phenols are readily brominated with elemental bromine and the resulting bromophenols efficiently extracted from water by carbon tetrachloride. The bromination reaction for phenol itself is

The IR determination is made using a 1-cm or 5-cm near-IR cell by measuring the absorbance of the band at 2.84 μm which arises from the stretching of the OH group internally hydrogen-bonded to a bromine in the ortho position. The method, consequently, fails for ortho-disubstituted phenols since the necessary bromination step is blocked. This is not a serious shortcoming, since there are few commercially important ortho-disubstituted phenols. For this reason, the IR method

TABLE 10

Detection Limits for the IR Determination of Phenols in Water[a]

Sample volume (ml)	CCl$_4$ volume (ml)	Minimum detection limit (ppm)
1000	20	0.01
1000	60	0.03
1000	100	0.05
2000	20	0.005
2000	60	0.015
2000	100	0.025
3000	20	0.003
3000	60	0.01
3000	100	0.02

[a]Measured with Wilks Scientific Miran-I Analyzer, cell path 5 cm.

is considered to be superior to the u.v. method, which fails for the more important para-substituted phenols. Table 10 gives typical sensitivities for the IR method.

B. Floating Films

A wide variety of floating films are found on natural bodies of water. These include oil slicks from industrial sources and natural seepage, inorganic silt, contamination from municipal sewage systems, industrial wastes, and the like. Qualitative identification is the first step in assessing the potential harmfulness of these films and in identifying their source, so that remedial action may be taken when necessary. Such films tend to be very difficult to collect by bulk sampling methods and equally difficult to separate from the water once collected. Two methods have been devised which collect only the floating film and thereby solve the entire sampling problem at the outset.

Brown et al. [26] have found that oil slicks can be sampled using aluminum foil fastened to some rigid support. The supported foil is dipped into the oil slick and the oil is found to adhere to the aluminum while the water does not. For spectral measurements, the foil is set up as a mirror in an auxiliary optical system placed in the spectrometer sample compartment. The authors point out that the resulting spectrum is a transmission spectrum since the film is relatively thick and is simply traversed twice by the light beam. The spectra of samples taken in this way show far less water absorption than similar samples taken on transmission windows. The authors propose that the separation is due to a greater wetting of aluminum foil by oil than by water.

Baier [27] has shown that an internal reflection crystal, when dipped edgewise through a film on water, picks up the film as it is withdrawn. Moreover, the film deposited on the plate is the same thickness as the original film. If the plate is dipped repeatedly, multiple layers are built up. The technique is applicable to films which vary from monolayers to thick films. Figure 21 shows spectra of oil films produced and sampled in the laboratory by this technique.

1. Oil Fingerprinting

An important aspect of floating film analysis is the problem of "fingerprinting" oil samples in order to identify properly the source of an oil slick. Using a small spectrometer, Mattson et al. [28] showed that careful comparison of the intensities of absorption bands would allow differentiation between oil from natural seepages and that from oil spills.

Brown et al. [29], using a somewhat higher performance spectrometer, have found that oils can be differentiated by comparing the bands in the 1000- to 600-cm^{-1} region very carefully. Using this method, these authors have been able to differentiate between oil slicks from different sources on Narragansett Bay.

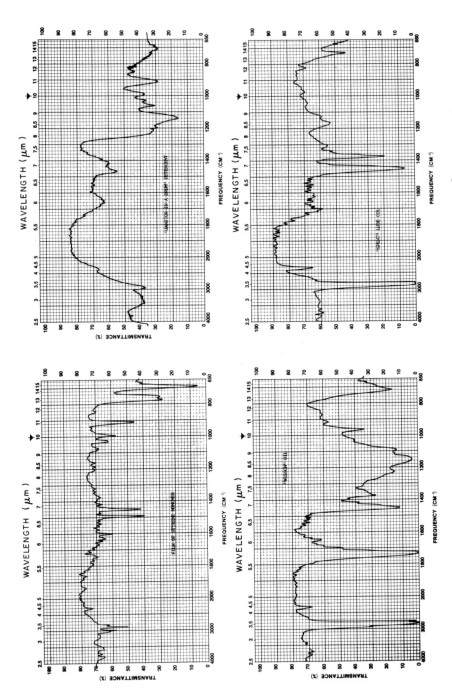

FIG. 21. Spectra of surface films on water sampled by dipping an ATR plate. Left: top, styrene monomer; bottom, "Wesson" oil. Right: top, "Janitor in a Drum" detergent; bottom, spent lube oil.

C. Inorganic Ions

Brown and Baldwin [30] have shown that the common inorganic ions SO_4^{2-}, NO_3^-, CO_3^{2-}, and PO_4^{3-} can be determined by laser Raman spectroscopy. Moreover, it is possible to determine these ions in water samples placed some distance from the spectrometer by using an auxiliary optical system to transfer the laser beam to the sample and the Raman scattering back to the spectrometer. Detection limits for normal sampling were NO_3^-, 25 ppm; PO_4^{3-}, 50 ppm; SO_4^{2-}, 50 ppm, and CO_3^{2-}, 75 ppm. With the sample 21 ft from the spectrometer, 150 ppm of nitrate ion could be detected.

REFERENCES

1. J. U. White, *J. Opt. Soc. Am.*, *32*, 285 (1942).

2. Philip L. Hanst in *Advances in Environmental Science and Technology*, Vol. 2 (J. N. Pitts, Jr., and R. L. Metcalf, eds.), Wiley (Interscience), New York, 1971.

3. J. D. Ingle, Jr., *J. Chem. Ed.*, *51*, 100 (1974).

4. Gary O. Nelson, *Controlled Test Atmospheres*, Ann Arbor Science Publishers, Ann Arbor, 1971.

5. Morris Katz, ed., *Methods of Air Sampling and Analysis*, American Public Health Association, Washington, 1972.

6. Andrew E. O'Keeffe and Gordon C. Ortman, *Anal. Chem. 38*, 760 (1966).

7. J. Rud Nielson, V. Thornton, and E. Brock Dale, *Rev. Mod. Phys. 16*, 307 (1944).

8. John A. Jamieson, Raymond H. McFee, Gilbert N. Plass, Robert H. Grube, and Robert G. Richards, *Infrared Physics and Engineering*, McGraw-Hill, New York, 1963.

9. S. S. Penner and D. Weber, *J. Chem. Phys.*, *19*, 807 (1951).

10. Philip L. Hanst, Allen S. Lefohn, and Bruce W. Gay, Jr., *Appl. Spect. 27*, 188 (1973).

11. G. Herzberg, *Molecular Spectra and Molecular Structure I. Spectra of Diatomic Molecules*, Van Nostrand, New York, 1950, p. 126.

12. R. D. Stewart, Hugh C. Dodd, Harold H. Gay and Duncan S. Erley, *Arch Environ. Health, 20*, 64 (1970); R. D. Stewart, *J. Am. Medical Assoc., 208*, 1490 (1969); R. D. Stewart, E. A. Boettner, R. R. Southworth, and J. C. Cerny, *J. Am. Medical Assoc., 183*, 994 (1963); D. S. Erley and R. D. Stewart, *Arch. Internal Medicine, 111*, 656 (1963); R. D. Stewart and D. S. Erley, *J. Forensic Sci., 8*, 32 (1963); R. D. Stewart, H. C. Dodd, D. S. Erley, and

B. B. Holder, *J. Am. Medical Assoc., 193,* 1097 (1965); R. D. Stewart, T. N. Fisher, M. J. Hosko, E. D. Baretta, H. C. Dodd, and A. A. Herrmann, *Arch. Environ. Health, 26,* 1 (1973).

13. R. K. Ritchie and D. Kulawic, *Can. J. Spectr., 17,* 150 (1972).

14. J. Kertzman, ISA meeting, St. Louis, 1973.

15. R. D. Stewart and D. S. Erley, "Detection of Volatile Organic Compounds and Toxic Gases in Humans by Rapid Infrared Techniques," in *Progress in Chemical Toxicology,* Vol. 2 (A. Stolman, ed.), Academic Press, New York, 1965, pp. 183-220.

16. B. B. Baker, E. I. du Pont de Nemours and Co., Wilmington, Delaware, to be published.

17. H. W. Jacobs, unpublished results.

18. B. M. Harney, D. H. McCrea, and A. J. Forney, *The Application of Remote Sensing to Air Pollution Detection and Measurement,* U. S. Bureau of Mines, Information Circular, 8577 (1973).

19. E. R. Schildkraut, *Am. Lab., December 1972,* p. 23.

20. L. R. Snowman and R. J. Gillmeister, Joint Conference on Sensing of Environmental Pollutants, Paper 71-1059, Palo Alto, November 1971.

21. E. D. Hinkley, *Appl. Phys. Letters, 16,* 351 (1970).

22. API Method 717-58.

23. J. Grasselli, Standard Oil of Ohio, private communication.

24. J. Puglish and M. Gruenfled, "Analytical Quality Control," *EPA Newsletter, 18,* 8 (1973).

25. R. G. Simard, I. Hasegawa, W. Gandaruk, and C. E. Headington, *Anal. Chem. 23,* 1384 (1951).

26. C. W. Brown, P. F. Lynch, and M. Ahmadjian, *Anal. Chem. 46,* 183 (1974).

27. R. E. Baier, *Surface Quality Assessment of Natural Bodies of Water,* Applied Physics Department, Cornell Aeronautical Laboratory, Buffalo, New York.

28. J. S. Mattson, H. B. Marks, Jr., R. L. Kolpack, and C. E. Schutt, *Anal. Chem., 42* (1970).

29. C. W. Brown, P. F. Lynch, and M. Ahmadjian, *Environ. Sci. and Tech. 7,* 1123 (1973).

30. C. W. Brown and S. F. Baldwin, *Water Res., 6,* 1601 (1972).

31. D. G. Murcray, T. G. Kyle, F. H. Murcray, and W. J. Williams, *J. Opt. Soc. Amer., 59,* 1131 (1969).

32. R. A. Schindler, AIAA Paper No. 71-1108, Joint Conference on Sensing of Environmental Pollutants, Palo Alto, 1971.

Chapter 8

FOOD INDUSTRY

A. Eskamani

Analytical Division
Research and Development Center
Standard Oil Company (Ohio)
Cleveland, Ohio

I.   INTRODUCTION 623

II.   DETERMINATION OF WATER IN FOOD 624

III.   ANALYSIS OF FOOD CARBOHYDRATES 628

IV.   ANALYSIS OF FOOD PROTEINS. 635

V.   APPLICATIONS TO FOOD LIPIDS. 638

VI.   IDENTIFICATION OF FOOD FLAVORS 651

VII.   ANALYSIS OF VITAMINS 657

VIII.   ANALYSIS OF ADDITIVES. 659

IX.   CONCLUSION 661

REFERENCES 662

I.   INTRODUCTION

The use of infrared (IR) spectroscopy in the field of food science
has been limited in spite of its wide application in other areas of
organic chemistry. This can be attributed primarily to the complex-
ities of sample preparation and presentation. Recent developments

have resolved some of these problems and IR spectroscopy is now
finding increasing use in the direct quantitative study and evaluation of various food components.

The application of IR spectroscopy to food chemistry began indirectly with the analysis of fatty acids in 1920. It was not until
1950 that IR spectroscopy was used directly in the identification of
food components, specifically lipids. Its utility in food applications was tied to the development of efficient solvent extraction
methods and separation techniques, such as thin layer chromatography,
paper chromatography, gas chromatography, and liquid chromatography.
IR spectroscopy, in combination with these separation techniques,
has become one of the most useful analytical tools in the field of
food science.

Raman spectroscopy, like IR, has been slow to find application
in the field of food science, but for different reasons. Its use
was restricted because of difficulties in obtaining spectra of small
samples. Recent advances in Raman instrumentation, especially newly
developed light excitation sources such as the laser beam, have eliminated some of the inherent problems.

In this chapter, applications of IR and Raman spectroscopy in
the identification of food components such as moisture, carbohydrates,
proteins, lipids, vitamins, flavors, and additives will be discussed.

The use of these techniques will enable food manufacturers to
establish quality control standards for evaluating incoming raw
materials, studying processing effects, and controlling quality of
finished products. These techniques will be vital to the food industry in meeting government regulations involving labeling of major
nutrients and additives which provide highly nutritious food with
extended storage capability.

II. DETERMINATION OF WATER IN FOOD

The water content of food is important to both food producers and
consumers. The water contained in food systems such as "instant"
powders, dried vegetables, and "mixes" influences their keeping quality. Water is also critical to the success of such food processing
procedures as baking and gelatin formation.

Because water plays such a vital role in so many food systems, it is necessary to determine it with speed and accuracy over a wide range of concentrations ($\sim$0.1 to 90%). Both physical and chemical methods have been reported [1-4]. The choice of method depends on the nature of the food, the amount of water present, and the precision required.

Conventional oven drying methods for water determination have several drawbacks. Loss of volatile components and chemical reactions producing water during the drying process contribute to the poor precision of these methods. The use of IR spectroscopy for water analysis overcomes many of these difficulties and, in many cases, requires less time.

Brandenberger and Bader [5] developed a method for determining the moisture content of instant coffee that has found wide application in other areas where moisture plays an important role in the keeping quality of the product. Moisture in the powder, removed by azeotropic distillation with dioxane, is determined by near-IR spectrophotometric analysis of the dioxane distillate. The method requires 20 min and has a reproducibility of ±0.02%.

Gold [6], Kay (cited in Ref. 6), and Hart et al. [7] have used near-IR to determine the water content of fruits and vegetables, cereal, and seeds, respectively. Their methods are based on an O-H frequency of water at 1.94 μm (5160 cm^{-1}). Water is removed from the samples by methanol extraction. The application of Hart's method to several foods is shown in Fig. 1. Table 1 shows that Gold's method correlates well with the vacuum-oven method over a wide range of water concentration (2 to 99%).

A collaborative study, reported by Rader [8], compares the near-IR method with the Karl Fischer Volumetric method and the vacuum-oven method for analyzing moisture in dried vegetables and spices. The near-IR method, applied to such vegetables and spices as green onion, garlic, toasted onion, and peas, was as accurate as and more rapid and specific than the vacuum-oven method. This method was recommended for adoption as the official method of the Association of Official Agricultural Chemists (AOAC) [9].

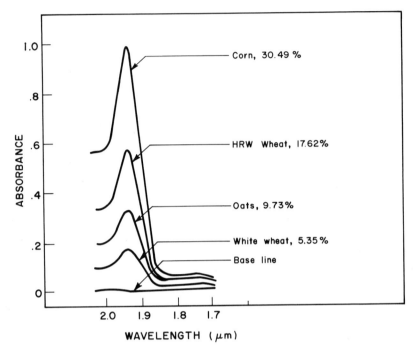

FIG. 1. Spectrograms of methanol extracts of grain with specified
moisture content. (Reprinted from Ref. 7.)

Near-IR absorption has been successfully applied to the determination of moisture in dairy products [10]. Methanol is used as the
extraction solvent due to its high affinity for water and unique
ability to remove water of hydration from lactose. Goulden and
Manning [10] reported absorbance measurements, at the 1.93 µm water
band, on methanol extracts of butter, cheese, and dried milk. Coefficients of variation in this method were 1.0, 1.1, and 2.0% for
butter, cheese and dried milk, respectively. The near-IR analysis
of moisture in butter, cheese, and cheese curd requires only a few
minutes; however, prior sample preparation procedure for determining
moisture in dried milk requires 2 hr standing time to eliminate the

TABLE 1

Comparison of Moisture Values by the Vacuum-oven Method
and the Near-IR Method[a]

Material	% moisture	
	Oven method	Near-IR method
Brewed coffee	98.99	99.49
Celery juice	97.97	97.70
Celery juice	96.69	96.81
Blanched celery	95.66	95.86
Celery stalk	95.45	94.98
Celery tops	94.45	95.01
Cabbage	91.52	91.54
Carrots	90.38	90.36
Reconstituted dry celery	89.21	88.71
Reconstituted tangerine juice concentrate	87.29	87.88
Partially dried celery, rehydrated	84.33	85.29
Concentrated tangerine juice	57.61	57.44
Partially dried celery	32.61	32.22
Commercial cornstarch	12.39	13.01
Powdered orange juice	3.33	3.27
Dried celery	3.19	3.28
Sweet potato flakes	1.76	2.89

[a]Results are averages of duplicate determinations. (Reprinted from Ref. 6.)

difficulties associated with lactose precipitation. The Goulden
method correlated well with the oven drying method for commercial
butter and margarine samples. The author has suggested the applicability of this IR method to process control in cheese-making; the
moisture content of cheese curd from the vat could be determined in
less than 5 min.

The water content of meat products poses a special problem for
meat processors and packers because of its regulation by the federal
government. The necessity for meat processors to comply with the
requirements established by the government has brought about the need
for a rapid and accurate method of moisture analysis. Ben-Gera and
Norris [11], taking advantage of the previously published literature
on near-IR spectroscopy, reported a unique application in the direct
determination of moisture in meat products. Near-IR spectral absorptions of meat emulsions were measured directly using a special
sample holder of 2 mm thickness. The absorbance measured at 1.80 μm,
subtracting the contribution from fat C-H at 1.725 μm, gives a high
correlation with the moisture content of meat. The calculated standard error was ±1.4%. The method was explored for many meat products
such as ham, bacon, and frankfurters.

Eisenbrand and Baumann [12] have developed a method for the
rapid determination of water in beer and other beverages based on
the near-IR absorption band of water at 0.977 μm. Their method has
made it possible to measure water and alcohol beverages simultaneously.

III. ANALYSIS OF FOOD CARBOHYDRATES

In those areas of food chemistry which deal with compounds of similar
chemical and physical properties, IR spectroscopy is an indispensable
tool for their analysis and identification. Such is the case in food
carbohydrate chemistry where the components of mixtures of chemically
similar sugars and starches must be identified.

The importance of IR spectroscopy in carbohydrate chemistry has
been recognized for some time. As early as 1950, Kuhn [13] obtained
IR absorption curves for a number of sugars, sugar derivatives, and

cellulosic materials. Anomeric forms of various glycosides, sugar esters, isomeric sugars, and sugar alcohols were readily distinguished by their IR curves. Selected IR spectra of sugars commonly encountered in food systems are presented in Fig. 2.

Whistler and House [14] have examined the IR absorptions of a number of monosaccharides and their derivatives and suggested that certain regions of absorption were characteristic of the configuration of the anomeric carbon and could be used to identify its α or β structure. For example, in the 9.4- to 9.8-μm region the α-isomer of most monosaccharides and their derivatives absorbs at a higher wavelength than the corresponding anomeric β form. The publication contains a useful table of IR analytical absorption values for many sugars and their derivatives. Similar data reported by White et al. [15] on the IR spectra of ten common disaccharides revealed significant spectral differences which allowed closely related disaccharides to be differentiated.

Goulden [16] was one of the first to use IR spectroscopy for the quantitative analysis of milk sugar (lactose). Under suitable conditions, the IR spectra of homogenized milk samples showed an absorption peak near 9.6 μm for lactose. Absorbance measured at 9.6 μm was directly proportional to the concentration of lactose in milk. Later, in a collaborative study reported by Biggs [17], it was suggested to the AOAC that the IR spectrophotometric milk analysis method be considered as the official method for lactose measurement in milk. Five different laboratories compared the IR method with the standard method for lactose analysis and obtained good agreement as shown in Table 2.

Because of the increasing use of pectic substances in food and pharmaceuticals and the commercial appearance of several modifications of pectin, its identification and measurement have become essential. Reintjis et al. [18] obtained IR spectra on a number of commercially available pectic substances and observed it was possible to distinguish between pectins with high and low methoxyl content. This information can be utilized for process control in making jelly, jams,

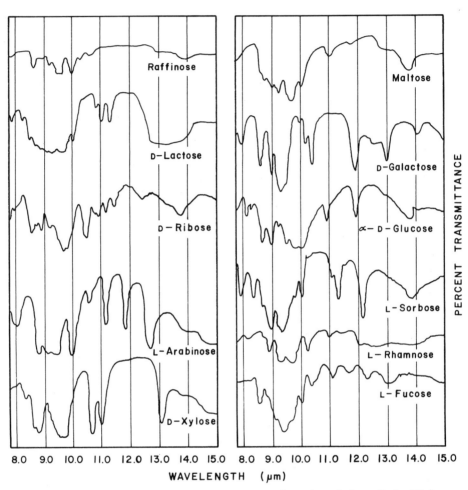

FIG. 2.   IR spectra of carbohydrates.   (Reprinted from Ref. 13.)

and preserves.   In the jelly-making process, the firmness of the gel
depends on factors such as percent pectin, molecular weight of the
pectins, percent methylation, percent sugar, and, finally, pH.   One
of the crucial parameters in the production of firm gel is the degree
of esterification of the carboxyl groups of the pectinic acid.   It is
well known that esterification of half of the carboxyl groups results

in excellent gelation [19]. Thus IR spectroscopy could be used as a quality and process control tool in the gel formation process.

Binkley et al. [20] have utilized IR spectroscopy to identify reducing sugars in the final molasses during the production of sucrose from cane juice. The identification of reducing sugars, largely D-glucose and D-fructose, is a laborious process by classical methods. The authors have pointed out that the "fingerprint" region (1000 to 700 cm^{-1}) of the IR spectrum of each reducing sugar is characteristic and different. By taking advantage of this difference, they developed a facile and rapid method for their identification. Thus, IR spectroscopy can be used to monitor reactions occurring during the thermal production of sucrose from cane juice.

In the late 1960s and early 1970s, many researchers [21-25] studied attenuated total reflection (ATR) as a possible rapid technique for quantitatively measuring major components of food materials. By using models of food systems, they were able to measure percent carbohydrate (starch), fat, and protein. Starch showed an intense absorption band at 990 cm^{-1}, which could be used to measure quantitatively its concentration. Later these researchers developed a novel sample holder which further extended the applicability of the ATR technique. ATR has also been used to quantitatively measure the carbohydrate content of soy products. An ATR spectrum of soybean starch is shown in Fig. 3. The strong broad band at 1040 cm^{-1} in the carbohydrate was assigned to C-C and C-O vibrations. It should be noted that C-C and C-O vibrations also exist in fat and protein; thus carbohydrate quantification by ATR requires differential analyses. Somewhat better IR characterization data were reported by Lin and Pomeranz [26] on the major components of wheat flour, especially wheat starch. The IR spectra of starch in wheat flour showed three prominent hydrogen stretching bands near 3000 cm^{-1}, a number of double bond stretching bands in the range 1875 to 1475 cm^{-1}, and typical type I ring stretching and type III vibrations at 920 and 770 cm^{-1}, respectively.

Masakazu and co-workers [27] have obtained far-IR spectra for glucose and saccharose, but this region of the IR has not found extensive application in food systems.

TABLE 2

Comparison of Results by IRMA and Standard Methods, Collaborative Studies[a]

Sample no.	ODAF		U. of G.		BCDA		Ross Lab.		Dairyman's	
	Std	IRMA	Std	IRMA	Std	IRMA	Std	IRMA	Std	IRMA
					Fat, %					
1	3.40	3.40	3.40	3.42	–	3.40	3.40	3.40	3.40	3.42
2	5.06	5.07	5.07	5.06	–	5.06	5.09	5.11	5.10	5.08
3	3.26	3.30	3.24	3.28	–	3.26	3.25	3.31	3.35	3.33
4	2.67	2.66	2.65	2.64	–	2.68	2.67	2.67	2.70	2.71
5	4.20	4.19	4.20	4.20	–	4.20	4.21	4.20	4.21	4.21
6	3.41	–	3.41	3.44	3.41	3.44	3.37	3.56	3.43	3.44
7	5.10	–	5.15	5.16	5.15	5.17	5.12	5.27	5.15	5.16
8	3.26	–	3.28	3.26	3.28	3.26	3.30	3.38	3.33	3.28
9	2.67	–	2.67	2.64	2.67	2.63	2.65	2.71	2.65	2.64
10	4.22	–	4.20	4.20	4.20	4.19	4.33	4.35	4.27	4.25
Mean	3.725	3.724	3.727	3.730	3.742	3.729	3.739	3.796	3.759	3.752
					Protein, %					
1	3.25	3.27	3.27	3.26	–	3.25	3.20	3.14	–	3.18
2	3.83	3.88	3.88	3.83	–	3.83	3.70	3.75	–	3.68
3	3.14	3.14	3.19	3.11	–	3.15	3.00	2.94	–	3.01
4	3.10	3.18	3.15	3.14	–	3.13	3.03	3.02	–	2.98
5	3.49	3.53	3.57	3.48	–	3.49	3.38	3.39	–	3.38

	C1	C2	C3	C4	C5	C6	C7	C8	C9	C10
6	3.26	–	3.26	3.30	3.15	3.29	3.20	3.18	–	3.18
7	3.80	–	3.86	3.89	3.75	3.81	3.70	3.87	–	3.83
8	3.14	–	3.17	3.17	3.08	3.16	2.99	3.03	–	3.05
9	3.13	–	3.13	3.19	3.09	3.15	3.05	3.05	–	3.01
10	3.52	–	3.53	3.58	3.45	3.53	3.38	3.49	–	3.46
Mean	3.366	3.400	3.401	3.395	3.307	3.379	3.263	3.286	–	3.276

Lactose, %

	C1	C2	C3	C4	C5	C6	C7	C8	C9	C10
1	4.57	4.60	4.66	4.59	4.57	4.57	–	4.82	–	4.69
2	4.56	4.54	4.68	4.51	–	4.55	–	4.74	–	4.67
3	4.58	4.62	4.69	4.59	–	4.58	–	4.82	–	4.72
4	4.48	4.58	4.58	4.56	–	4.49	–	4.76	–	4.65
5	4.59	4.60	4.71	4.56	–	4.60	–	4.80	–	4.66
6	4.55	–	4.65	4.61	4.60	4.56	–	4.90	–	4.81
7	4.58	–	4.68	4.59	4.65	4.59	–	4.98	–	4.84
8	4.60	–	4.69	4.65	4.67	4.59	–	4.93	–	4.83
9	4.48	–	4.59	4.61	4.52	4.49	–	4.84	–	4.74
10	4.63	–	4.72	4.66	4.65	4.65	–	5.00	–	4.89
Mean	4.562	4.588	4.665	4.593	4.618	4.567	–	4.859	–	4.750

[a]ODAF = Ontario Department of Agriculture and Food; U. of G. = University of Guelph; BCDA = British Columbia Department of Agriculture. (Reprinted from Ref. 17.)

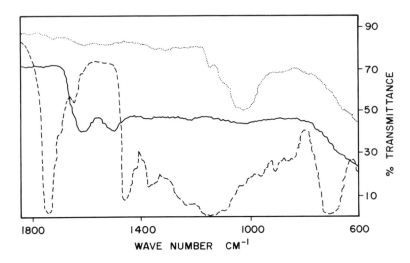

FIG. 3.   ATR spectra of 100% soybean oil, protein, and carbohydrate.
.... = carbohydrate; ───── = protein; ─ ─ ─ ─ = fat.   (Reprinted
from Ref. 25.)

Recent developments have occurred in the area of low-temperature
IR spectroscopy of carbohydrates.   The effect of temperature on spectral quality is impressive.   Michell [28] has published a series of
IR spectra on naturally occurring polysaccharides which show a marked
increase in spectral definition at lower temperatures.   A more in-
depth study on a large number of carbohydrates having IR spectra
ranging from 2000 to 250 cm^{-1} at 110 K and 4000 to 33 cm^{-1} at 20 K
was reported by Katon et al. [29].   The improved spectral quality of
carbohydrates at lower temperatures is associated with the strong
intermolecular hydrogen bonds of the hydroxyl groups.

One major drawback in the use of IR spectroscopy for carbohydrate identification is the interference of the strong water absorption band.   The recent introduction of Raman spectroscopy, especially
laser Raman, has offered a solution to this problem.   Vasko and co-
workers [30] have published many excellent articles on the use of
IR and Raman spectroscopy in carbohydrate analysis.   Raman spectroscopy has made it possible to obtain spectra for carbohydrates in
water solution since water is not an intense Raman scatterer.   This
makes Raman spectroscopy an ideal technique for the study of food

carbohydrates in aqueous solutions. Raman spectroscopy also has
another advantage over IR, namely, the strong absorption shown by
C=N groups [31]. The combination of laser Raman and IR spectroscopy
could be a powerful tool for the analysis of food carbohydrates. In
particular, Raman spectroscopy would be applicable for the analysis
of the sugar content of beverages.

IV. ANALYSIS OF FOOD PROTEINS

Applications of IR spectroscopy in the field of food protein have
been limited due to the availability of relatively rapid chemical
methods for analyzing proteins. However, IR spectroscopic methods
have recently begun to penetrate the field of food protein chemistry.
Lin and co-workers characterized the protein in wheat flour by IR
spectroscopy [26]. The protein was measured by amide I, II, III,
and IV absorptions at 1650, 1550, 1270, and 675 cm^{-1}, respectively.
In addition, C-H stretching and CH_2 and CH_3 deformation data were
utilized. The authors were able to relate the IR spectra with the
functional properties of flour. IR spectra of five different flour
classes, such as hard red winter, hard red spring, soft red winter,
durum, and white club wheats which vary widely in protein composition
and bread-making potential were obtained. The method could be used
to evaluate the physicochemical properties of various wheat flours.

   The analysis of protein in milk by IR spectroscopy has received
more emphasis than the analysis of proteins in other foods mainly
because of ease of sampling. Kliman and Pallansch [21] were the
first to formally report data on the ATR spectra of milk and their
quantitative relationship to its composition. The broad and general
band in the 1035-cm^{-1} region was found to be directly proportional
to the total solids in whole milk or skim milk. The authors have
stated that the linear relationship between solids content and ATR
band absorption could be used to monitor total solids in the process
of high-speed milk evaporation.

   One of the early applications of IR spectroscopy in the milk
industry was the Infrared Milk Analyzer (IRMA) introduced in England
by Goulden (cited in Ref. 22). IRMA could measure milk protein in

33 sec. The instrument offered another outstanding advantage in
that it did not require any chemicals for sample preparation. In
1970 and 1972 Weik and Biggs [17,24] suggested that the IRMA technique become the official method for milk protein analysis. Later
in 1972, the proposed IRMA method was compared with the standard
method used by various laboratories and results were in excellent
agreement. The data are illustrated in Table 2. The ATR method
has been applied successfully to the analysis of soy protein and
other food systems by Wilson and co-workers [23,25]. Using special
baseline techniques, it was possible to quantitatively determine
protein content of soy products. Analytical wavelengths assigned to
protein were absorption bands in the transmission spectrum of amide
I at 1660 to 1640 cm^{-1} (C=O stretching) and amide II at 1570 to 1350
cm^{-1} (N-H in-plane bending and C-N stretching) as shown in Fig. 3.

Wheat proteins such as glutenin and gliadin are vital to the
baking process [19]. The composition of these proteins has been
closely examined by IR spectroscopy. IR studies have revealed that
gliadin consists chiefly of an α-helical and unordered structure [32].
Further studies on wheat protein by Wehrli and Pomeranz [33] provided
a better understanding of the gliadin structure. IR spectroscopy
has been utilized to study the interaction of protein and glycolipids
in wheat flour. It has been one of the few techniques to reveal
information on protein conformation and interactions. The gliadin-
lipid-glutenin interactions are important since they determine the
functional properties of wheat flour, namely, its bread-making properties.

The relationship between water and wheat flour is of great
interest to the bread-making industry. In the presence of water,
and with mechanical agitation, wheat flour forms a tough, elastic
complex called gluten. Gluten is capable of retaining gases. Knowledge about the physiochemical state of water surrounding the wheat
flour should shed some light on the dough formation mechanism.
Hopefully, IR analysis will be a useful tool in studying this mechanism. Buontempo et al. [34] studied the interaction of biopolymers

and water. Their IR data indicated that, in the case of polyamino-
acids, the water band was displaced from 3400 cm^{-1} to 3500 cm^{-1}.
This was attributed to the effect of water on the -NH band.

The feasibility of using Raman spectroscopy to obtain valuable
information on proteins has been demonstrated in the literature [35-
37]. Recently, laser Raman spectroscopy has been successfully employed to tabulate the active vibrations of various proteins. Basically, Raman and IR spectroscopy provide the same kind of information.
However, Raman has one important advantage over IR, namely, its applicability to aqueous samples with little interference by the solvent. This advantage permits spectroscopic study of systems that
are very difficult to investigate by IR spectroscopy. The Raman
spectrum of water has only one moderately intense band in the region
3200 to 3600 cm^{-1} and weak bands near 1640, 800, 450, and 75 cm^{-1}.
The recent development of the laser will provide an ideal source of
excitation of the Raman effect in biological systems such as food.

The study of protein denaturation has been one of the most important and interesting applications of Raman spectroscopy in food
chemistry. Brown [37] reported Raman bands lower than 50 cm^{-1} in
proteins. His study showed a peak at 29 om^{-1} for proteins except
those that were denatured. Similarly, Lord and Yu [35] have strongly
suggested the usefulness of laser-excited Raman spectra in establishing the presence and number of disulfide cross-links in proteins
(C-S-S-C). The peptide CONH groups which give rise to two characteristic lines near 1660 cm^{-1}, due to amide I, and 1260 cm^{-1}, due to
amide III, appear to be potentially useful in assessing conformational
changes caused by denaturation. Also, lines in the region from 800
to 1150 cm^{-1} due to C-C and C-N stretching vibrations have been related to conformation changes. The above information would be extremely useful in food protein study. Many food processes such as
canning, dehydration, sterilization, and freeze-drying greatly alter
the food protein structure. Denatured food protein has low nutritional value and solubility problems. Therefore, IR and especially
Raman spectroscopy will become ideal tools for the study of food

protein. Raman spectroscopy will also provide extremely useful information on the properties and characteristics of texturized food
proteins.

V. APPLICATIONS TO FOOD LIPIDS

Although IR spectra were known to offer a great deal of information
on specific vibrating groups, little application was made of this
analytical tool in the analysis of fats and oils. It was not until
1920 that Gibson (cited in Ref. 38) published a paper exclusively
devoted to the application of IR spectroscopy to fatty acid materials.
The more widespread application of IR spectroscopy to the analysis
of oils and fats occurred during the period 1940 to 1950. Since 1950,
IR spectroscopy has become an indispensable tool for lipid chemists.

Extensive IR spectroscopic data were reported by Shreve et al.
[39] on many long-chain fatty acids, esters, and alcohols. The above
classes of compounds were readily characterized on the basis of spectral differences common to each class. Spectral data were utilized
to differentiate cis/trans and saturated/unsaturated compounds with
no difficulty. In 1956, O'Connor [38] published an excellent review
on the application of IR spectroscopy to fatty acid derivatives.
His review included areas of interest to food lipid chemists, namely,
cis/trans isomerization, auto-oxidation, and rancidity. Spectral
data compiled by O'Connor are shown in Table 3. Other review articles
on the IR spectroscopy of fats and oils are by Davenport [40], Freeman
[41] and others [42-48].

It was not until the introduction of improved sample preparation
and separation techniques that IR spectroscopy became a popular
identification tool in food chemistry [49]. Lipids can be conveniently examined as films, solutions, mulls, or in alkali halide pellets
[40]. If the lipid is liquid at room temperature, its spectrum can
be determined as a film pressed between two rock salt plates. If it
is a solid, a mull can be made by grinding it with a paraffin oil
(Nujol).

TABLE 3

One Hundred Absorption Bands Employed in the Applications
of IR Spectroscopy to Fatty Acid Chemistry[a]

No.	Wavelength position of observed absorption band (μm)	Vibrating group giving rise to observed absorption band
		I. O-H, C-H, N-H, C-D, P-OH, and P-H stretching vibrations. Regions 2 to 5 μm
		A. O-H stretching
1	2.75-2.80	Free-O-H
2	2.82-2.90	Bonded-O-H$\cdots$O of single-bridged dimer
3	2.95-3.25	Bonded H$\cdots$O-H$\cdots$O- of double-bridged polymer or cyclic
		B. C-H stretching
4	3.00-3.05	R$\equiv$C-H
5	3.22-3.25	R_2=C-H_2
6	3.28-3.32	R_2=C-HR
7	3.40-3.45	R-C-H_3
8	3.42-3.50	R_2-C-H_2
9	3.45-3.48	R_3-CH
10	3.50-3.70	R-C-H
11	3.70	C-H and bonded O-H$\cdots$O combination band
		C. N-H stretching
12	2.85 and 2.95	Free N-H primary amide
13	3.00 and 3.15	Bonded N-H$\cdots$primary amide
14	2.90-2.95	Free N-H secondary amide

TABLE 3 (continued)

No.	Wavelength position of observed absorption band (μm)	Vibrating group giving rise to observed absorption band
15	3.00-3.05	Bonded N-H⋯secondary amide, single bridge (trans)
16.	3.15-3.18	Bonded N-H⋯secondary amide, single bridge (cis)
17	3.22-3.25	Bonded N-H⋯secondary amide cyclic dimer
18	2.85 and 3.02	N-H primary amine
19	2.85-3.02	N-H secondary amine
20	2.95-3.12	N-H imines
21	3.20-3.30	$N-H_3^+$ amino acids
		D. C-D stretching
22	4.64	C-D
23	4.84	C-D
		E. P-H, P-OH stretchings
24	4.05-4.25	P-H
25	3.70-3.90	P-OH

II. C=O and C=C, C≡C stretching vibrations. Region 3 to 6 μm

A. Aldehydes

26	5.75-5.80	$\overset{\displaystyle H}{\underset{\displaystyle \diagup}{}}$ RC=O, saturated
27	5.83-5.90	$\overset{\displaystyle H}{\underset{\displaystyle \diagup}{}}$ PhC=O, aryl
28	5.85-5.95	$\overset{\displaystyle H}{\underset{\displaystyle \diagup}{}}$ R-CH=CH-C=O, α,β unsaturated

B. Ketones

29	5.80-5.85	$RCH_2-\overset{O}{\overset{\|\|}{C}}-CH_2R$, saturated

TABLE 3 (continued)

No.	Wavelength position of observed absorption band (μm)	Vibrating group giving rise to observed absorption band
30	5.90-5.95	Ph-C(=O)-CH$_3$, aryl-alkyl
31	6.00-6.02	Ph-C(=O)-Ph, diaryl
32	6.00-6.05	R-CH=CH-C(=O)-R, α,β unsaturated
33	5.63	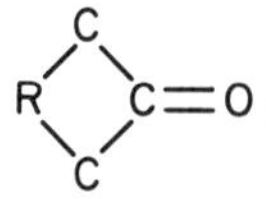 C=O, 4-membered, saturated ring
34	5.73	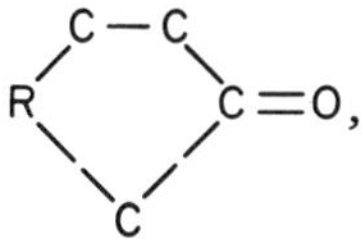 C=O, 5-membered, saturated ring
35	5.81	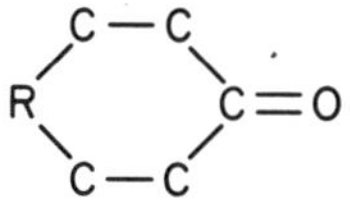 C=O, 6-membered, saturated ring, or C=O, 5-membered, α,β unsaturated ring
36	5.95	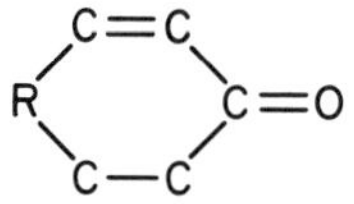 C=O, 6- (or 7-) membered, α,β unsaturated ring
37	5.90-6.00	O=Ph=O, quinone, 2C=O's on 1 ring
38	6.05-6.10	O=Ph-Ph=O, quinone, 2C=O's on 2 rings

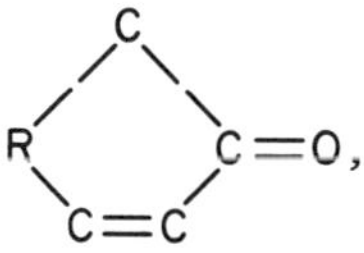

C. Acids

No.	Wavelength position	Vibrating group
39	5.68	R-C(=O)-OH, saturated monomer

TABLE 3 (continued)

No.	Wavelength position of observed absorption band (μm)	Vibrating group giving rise to observed absorption band
40	5.80-5.88	O···HO // R-C-OH···O=C-R, saturated dimer
41	5.90-5.92	OH / R-C=C-C=O, α,β unsaturated
42	5.90-5.95	OH / Ph-C=O, aryl
43	6.00-6.05	Chelated hydroxy acids, some dicarboxylic acids

D. Esters

44	5.65	OCH₃ / H_2-C=C-C=O, vinyl ester
45	5.75	OCH₃ / R-C=O, saturated
46	5.80-5.82	OCH₃ / R-C=C-C=O, α,β unsaturated, or R-COOPh, aryl

E. Lactones

47	5.50	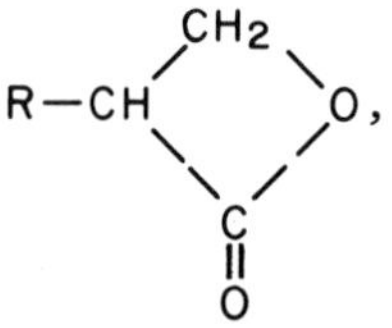 β, or 4-membered saturated ring
48	5.65	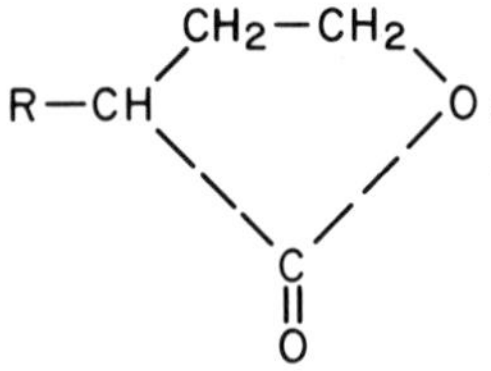 γ, or 5-membered saturated ring

TABLE 3 (continued)

No.	Wavelength position of observed absorption band (μm)	Vibrating group giving rise to observed absorption band
49	5.72	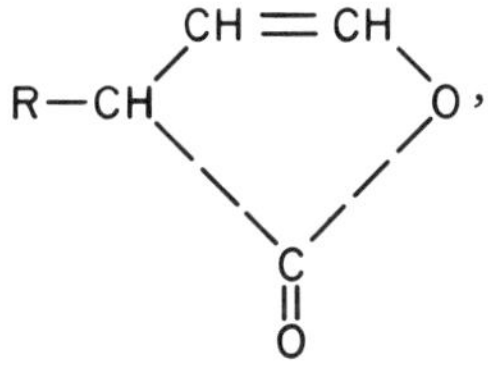 γ, or 5 membered α,β unsaturated ring
50	5.75	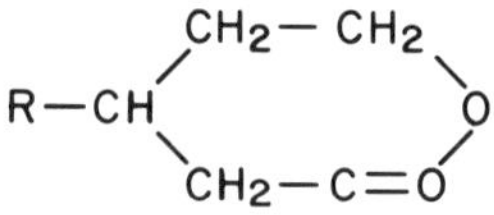 ε, or 6-membered saturated ring

F. C=C, C≡C stretching

51	6.0-6.1	C=C *cis* only (weak when internal in symmetrical molecules)
52	3.03	HC≡CH
53	4.67-4.76	RC≡CH
54	4.44-4.58	RC≡CR
55	5.14 and 9.45	C=C=C

III. C-H deformations, saturated groups. Region 6 to 7 μm

56	6.7-6.9	-C-H_2- group
57	6.8-7.0	-C-C-H_3 group, asymmetrical deformation
58	7.15-7.20	-C-$(C-H_3)_3$ group
59	7.20-7.25	-C-$(C-H_3)_2$ group
60	7.25-7.30	C-C-H_3 group symmetrical deformation
61	7.30-7.35	-C-$(C-H_3)_2$ group
62	7.45-7.50	-C-H- group

IV. C-O stretching and C-OH bending. Region 7.7 to 10.0 μm

TABLE 3 (continued)

No.	Wavelength position of observed absorption band (μm)	Vibrating group giving rise to observed absorption band
		A. Alcohols
63	7.2-8.6	Phenols
64	8.3-8.9	Tertiary open-chain saturated
65	8.9-9.2	Secondary open-chain saturated
66	9.2-9.5	Primary open-chain saturated
67	8.3-8.9	Highly symmetrically-branched secondary
68	8.9-9.2	α-Unsaturated or cyclic tertiary
69	9.1-9.2	Secondary with branching on one α-carbon
70	9.2-9.5	Secondary, α-unsaturated or alicyclic 5- or 6-membered ring
71	9.5-10.0	Secondary: di-unsaturated, α-branched and unsaturated, or 7- or 8-membered ring
		Primary: α-branched and/or unsaturated
		Tertiary: highly unsaturated
		B. Acids
72	7.75-7.80	C-O
73	8.40-8.45	C-O
		C. Esters
74	7.90-8.00	C-O
75	8.40-8.50	C-O
		D. Ethers
76	8.7-9.4	CH_2-O-CH_2, alkyl
77	7.8-8.1	Ph-O-Ph, aryl or =C-O, unsaturated
		E. Anhydrides
78	7.7-8.3	Cyclic
79	8.5-9.5	Open chain
		F. Phosphorus
80	6.90 and 10.0	P-O-Ph, aromatic
81	9.52	P-O-CH_3 aliphatic

TABLE 3 (continued)

No.	Wavelength position of observed absorption band (μm)	Vibrating group giving rise to observed absorption band
		V. C-H deformation about a C=C and skeletal and "breathing" vibrations. Region 10.0 to 15.0 μm
		A. C-H bending
82	10.05-10.15	X\C=C/H, H/ \H
83	10.20-10.36	X\C=C/H (trans only[b]), H/ \Y
84	10.90-11.05	H\C=C/H, X/ \H
85	11.17-11.30	X\C=C/H, Y/ \H
86	11.90-12.50	X\C=C/H, Y/ \Z
87	13.0>15.0	X\C=C/Y (cis only[b]), H/ \H
		B. Skeletal and "breathing"
88	9.75 and 11.55	Cyclopropane
89	10.9 and 11.3	Cyclobutane
90	10.31 and 11.16	Cyclopentane
91	9.63, 9.86 11.05 and 11.60	Cyclohexane

TABLE 3 (continued)

No.	Wavelength position of observed absorption band (μm)	Vibrating group giving rise to observed absorption band
92	11.2	Epoxy-oxirane ring derived from internal R-C=C-R (trans only)
93	12.0	Epoxy-oxirane ring derived from internal R-C=C-R (cis only)
94	11.8, 12.9, and 14.7	Benzene ring
95	12.0	Hydroperoxide
96	12.95	Ethyl
97	13.0	CH_2 rocking on long carbon chain
98	13.5	n-Propyl
99	13.8	Hydroperoxide
100	7.5-8.5	Progression of bands in solid state spectra, probably due to wagging and/or bending mode of vibration of the C-H bonds of methylene groups. The number of bands in the progression is indicative of chain length.

[a]The exact position of maximum absorption depends upon whether the measurements were made on the pure liquid, solid, mull, or solvent and on the nature of the particular solvent. Several band positions are also critically dependent upon neighboring groups.

The value and range given in this table are from the author's collection of these bands in fatty acid materials mostly from original reports in technical journals. They represent average values of ranges of the various data that have been reported for the specific absorption. (Reprinted from Ref. [38].)

[b]See text for table of wavelength positions of various combinations of conjugations involving these two internal groups.

One of the most useful applications of IR spectroscopy in food processing is the identification of cis/trans unsaturation. Most naturally occurring vegetable fats and unsaturated constituents contain only nonconjugated double bonds in the cis configuration. However, these cis bonds may be isomerized to the trans configuration

during extraction and processing procedures. Also, oxidation and
partial hydrogenation promote isomerization from the naturally occurring cis to the trans isomers. The IR method has been adopted
officially by the AOCS [50,51]. The method is based on an absorption
band with a maximum at 10.3 µm. The band arises from a C-H deformation about the trans double bond and is found in the spectra of all
compounds containing an isolated trans olefinic group. This band is
not observed in the spectra of the corresponding cis and saturated
compounds. The quantitative method for the determination of isolated trans content [52,53] is based on measurements of the intensity
of this absorption band. This is shown in Fig. 4.

    Cis/trans determination by IR spectroscopy is useful not only
for following polymerization, auto-oxidation, and hydrogenation, but
also for detecting adulturation of fat. Bartlett and Chapman [54]
used IR spectroscopy to determine hydrogenated (saturated) fats in

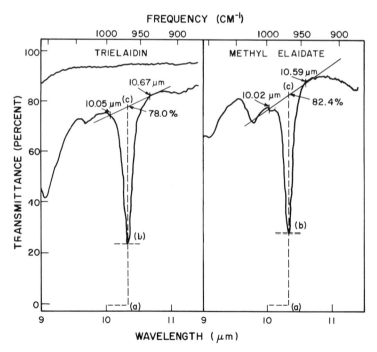

FIG. 4. IR absorption of trans unsaturation in esters.
(Reprinted from Ref. 51.)

butter by measuring cis/trans conjugated unsaturation. Butter contains cis/trans conjugated unsaturation as well as isolated trans
unsaturation while hydrogenated fats contain only the latter. Both
systems are detectable in the 920- to 990-cm^{-1} region of the spectrum. Conjugated and isolated unsaturation are present in a constant
ratio in pure butter; therefore, addition of hydrogenated fats greatly
increases the isolated trans double bonds (967 cm^{-1}) but leaves the
conjugated diene essentially unchanged (948 and 925 cm^{-1}). Thus,
it is possible to determine the percentage of a hydrogenated adulter-
ant fat using IR differential techniques. This is illustrated in
Fig. 5.

Cis/trans isomer composition data, especially in the case of
unsaturated esters in edible vegetable oils, provide valuable information on nutritional and storage quality. Thus, determination
of both cis and trans is important; however, IR spectroscopy can
determine trans more readily. Bailey [55] introduced Raman spectroscopy for the measurement of cis/trans isomer content of edible

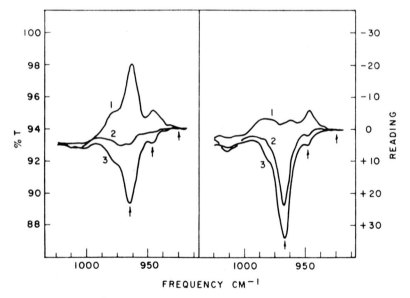

FIG. 5. Differential spectra. Left, pure butter; right, butter
containing 10% hydrogenated fat. (Reprinted from Ref. 54.)

vegetable oils. The intensities of Raman lines near 1656 and 1670 cm^{-1} are associated with cis and trans configuration and thus direct measurement of the amounts of both isomers is possible.

Studies of auto-oxidation and rancidity in relation to off-flavor production in foods cannot be overemphasized. Rancidity development in food products such as milk, meat, dehydrated food, potato chips, fruits, and vegetables is a prime concern of food chemists. The process of auto-oxidation and rancidity not only involves cis to trans isomerization but also production of many other compounds, such as peroxides and hydroperoxides. IR spectroscopy has been used successfully to follow the auto-oxidation process qualitatively and quantitatively. Bands at 2.93, 3.2, 5.72, and 10 to 11 μm have been associated quantitatively with hydroperoxides, free rancid, C=O, and cis to trans isomerization, respectively. Analysis by IR absorption was found to be more sensitive to changes than a qualified taste panel and considerably more sensitive than determination of peroxide values [38].

Arnold and Hartung [56] have successfully explored the utility of IR spectroscopy in the determination of degree of unsaturation of complex triglyceride mixtures found in food oils and fats. The ratio of absorbances of olefinic/saturated C-H stretching correlates (0.98) well with iodine number. The unsaturation in many food oils and fats such as corn, lard, milk, fat, peanut oil, soybean oil, vegetable oil, cottonseed oil, and many other oils can be determined by this technique. Some important IR absorption patterns for a few oils are illustrated in Fig. 6. The authors have developed a "prediction equation" which enables one to calculate iodine value from IR data as shown below:

Iodine value = 495 (IR ratio) - 54.1

$$IR \ ratio = \frac{absorbance_{3.3 \ \mu m}}{absorbance_{3.5 \ \mu m}}$$

This IR technique should have potential in the development of IR instruments specially designed to monitor unsaturation in process control applications.

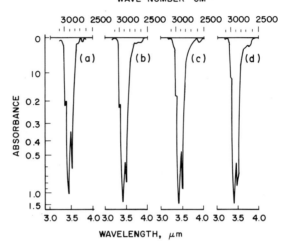

FIG. 6. IR absorption patterns in 3- to 4-μm region of (a) safflower oil, iodine value 144; (b) soybean oil, iodine value 109; (c) lard, iodine value 73; and (d) milk fat, iodine value 33. (Reprinted from Ref. 56.)

IR spectroscopy has been successfully applied to the quantitative measurement of milk fat. Automatic measurement of milk fat by IRMA (IR Milk Analyzer) has been practiced for several years [22,24]. Biggs [17] has reported IR data on milk fat and compared these with other methods of fat analysis conducted at various laboratories. Data in Table 2 show an excellent correlation among the various methods.

A study of ATR on milk products was published by Kliman and Pallansch [21]. Their ATR spectra of milk showed characteristic bands at 1735 cm^{-1} and 1035 cm^{-1} for fat and total solids, respectively. The best correlation with fat content was obtained at the 1735 cm^{-1} C=O stretching (Fig. 3). The ATR technique has been applied to the quantitative determination of fat in food systems such as soy products also [23,25].

In addition, near-IR spectroscopy has been utilized as a rapid and simple spectrophotometric technique for determining fat content of milk [57]. The method is based on the intensification of the

0.97-μm water absorption band resulting from the presence of fat in milk. The above approach has been applied to fat analysis of whole milk, skim milk, and relatively thick milk samples. Similarly, near-IR spectral absorption properties of meat products were explored by the above authors [11]. Absorbance at 1.725 μm due to C-H was highly correlatable with the fat content of meat. This method has been replacing conventional methods of fat measurement such as extraction techniques.

Lin and Pomeranz [26] utilized IR spectroscopy to characterize the fat component of wheat products. Strong absorption bands near 3330 cm^{-1} (O-H stretching), 3000 cm^{-1} (C-H stretching), and 1750 cm^{-1} (C=O stretching) were associated with fat content. Attempts were made to relate minor differences in the spectra with the functional properties of flour.

IR spectroscopy of fat has been suggested as a quality control measure of many food products. For example, it could be used as a qualitative measure of the pureness of commercial cocoa butter. IR spectroscopy can differentiate oil originating from cocoa seed, which is considered pure, from that of cocoa shell, which is a contaminant.

VI. IDENTIFICATION OF FOOD FLAVORS

The chemistry of food flavors is an entire field of food science in itself. Flavor characterization has been difficult mainly because substances responsible for food flavors and aroma constitute such a small part of the whole food system. Many of the compounds that must be identified and measured by flavor chemists are present in ppm or ppb. In addition, compounds that contribute to the overall food flavor profile are comprised of many different classes of organic compounds. Chemical analysis of the flavor components of foods is important to the food industry for the following reasons [58]: (1) to understand the mechanism of formation of flavor compounds so that desirable flavors can be formed and undesirable flavors retarded, (2) to synthesize flavors for the enhancement of existing flavors or to impart fresh flavor to processed food, (3) to provide sufficient information so that food products with highly desirable flavors can be grown by genetic manipulation, and (4) to compile information which can be used in establishing quality control procedures for the flavor of food products.

The study of food flavor generally involves three major steps: (1) isolation of the flavor from food, (2) fractionation of total flavor substances into individual components, and (3) chemical identification of collected micro amounts of each fraction.

During the past decade, the chemical study of food flavors has taken a giant step due to the development of microchemical methods of isolation and instrumental techniques of identification. Since the physical and chemical nature of foods are different, various isolation techniques had to be developed. A few of these creative systems are illustrated in Figs. 7-9 [58,59].

In general, volatile flavor compounds are stripped by vacuum distillation utilizing techniques such as the molecular still, flash evaporation, and vaporization [58,59]. The flavors are trapped in

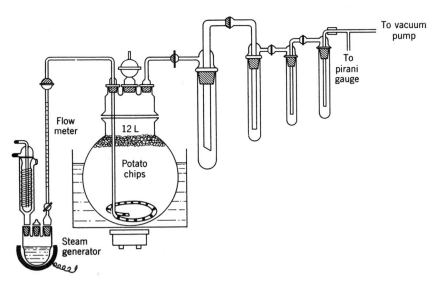

FIG. 7. Apparatus for the isolation of volatile flavor compounds from solid foods like potato chips. (Reprinted from Ref. 58.)

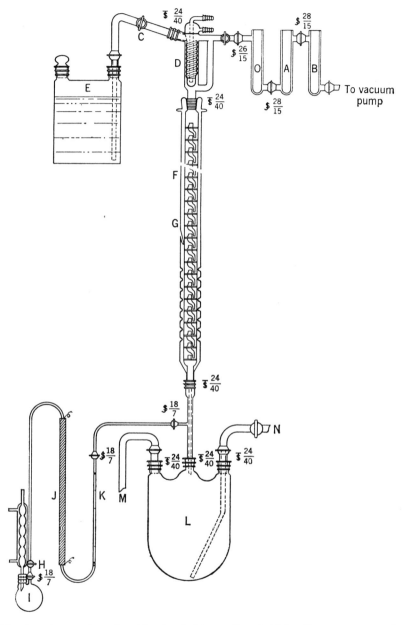

FIG. 8. Apparatus for the isolation of volatile flavor compounds from oils by semicontinuous countercurrent vacuum steam distillation. (Reprinted from Ref. 58.)

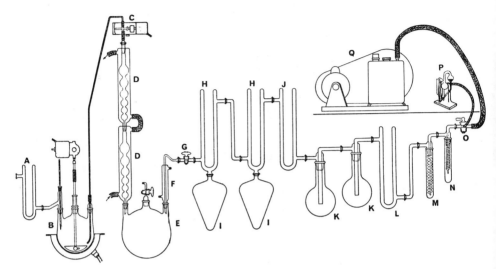

FIG. 9. Apparatus for the isolation of trace volatile constituents
from foods. (Reprinted from Ref. 59.)

cold fingers, fractionated by gas chromatography, and identified by
IR spectroscopy. IR spectroscopy has found one of its largest applications in food flavor chemistry. Some of its uses in the analysis
of food flavors will be discussed in the following paragraphs.

IR spectroscopy has been used extensively to identify citrus
flavors. Since the characteristic flavor and aroma of oranges are
associated with oil-soluble compounds, the volatile constituents of
the essential oil of the orange have been the subject of many studies
[60-72]. Two of the major components of the essential orange oil,
namely, terpenes and terpenoids, are comprised of 50 different organic compounds. Rymal et al. [73] utilized IR spectroscopy to
measure changes in the volatile flavor constituents of canned single-
strength orange juice due to storage temperature. His findings indicated quantitative reductions in d-limonene and linalool, and an
increase in furfural and α-terpineol in orange juices stored at 27°F
and 4°C. This illustrates a potential application of IR spectroscopy
for the quality control of orange juices.

IR spectroscopy has been applied successfully to the characterization of snack food flavors. One of the most widely studied food flavors is that of the potato chip. Chang et al. [58,74,75] conducted extensive research on potato chip flavor and identified a hundred general compounds as well as many unique ones. Their IR data have helped in the development of an authentic potato chip flavor.

Dornseifer and Powers [76] employed IR spectroscopy to study flavor changes in potato chips during storage. They found that high-molecular weight carbonyl and butyric acid were increased during storage and were associated with off-flavors in the product. These data suggest the usefulness of IR spectroscopy for quality control.

Off-flavors produced during milk processing and storage can be detected by IR spectroscopy [77-79]. Volatile milk compounds after heat treatment were identified and related to a "coconut-like" flavor [77]. The off-flavor was due to the presence of γ-dodecalactone. The compound was thought to have some significance in flavor changes of processed and stored dairy products [78]. Another off-flavor in milk known as "sunlight" flavor was identified as methional by IR data. Yet another off-flavor known as "phenolic" flavor was identified in Gouda cheeses by IR data as p-cresol and was attributed to the improper filtering of enzyme rennet. This technique showed the potential use of IR in process control [80].

A method for the identification and quantitative measurement of the constituents of maple flavor had long been needed. The information was needed to guide research on improving the quality of maple syrup, on developing syrup for specification studies, and finally for detecting adulteration. Underwood and Filipic [81] reported a method of extracting and identifying the flavor components of maple syrup. Extracted flavors, identified by IR spectra, were found to be mainly vanillin, syringaldehyde, and dihydroconiferyl alcohol.

The isolation and identification of meat flavors is still a difficult task and as yet incomplete, even though numerous advances have been made in experimental apparatus and instrumental analysis. IR spectroscopy has been indispensable in the identification of meat flavors and their precursors. Horstein and others [82-85] have made extensive flavor studies on beef, pork, lamb, and chicken. Careful analysis of IR spectral data has revealed that the lean portions of beef, lamb, and pork are responsible for the meaty flavor while characteristic differences in beef, lamb, and pork reside in the fat. Researchers [86,87] have identified some beef flavor precursors as low molecular weight peptides, carbohydrates, and phosphates. Hirai and co-workers [88] reported IR spectral data on 57 compounds asso- ciated with the boiled beef flavor. Their IR data revealed some interesting and usual compounds as major contributors to the overall boiled beef flavor.

The acceptability of dry cured country style ham (CS ham) is determined by odor and flavor; these economically important factors were studied by Ockerman [89]. His study related aging and the presence of carbonyl compounds.

IR spectroscopy has been employed for the identification of fruit and vegetable flavors. Forss et al. [90] identified cucumber flavor as nona-2,6-dienal, non-2-enal, hex-2-enal, and three aliphatic aldehydes. Further study showed that the pleasant odor of the cucum- ber was largely due to nona-2-trans, 6-cis-dienal with some contri- bution from hex-2-enal. The more unpleasant astringent flavor was contributed by non-2-enal.

IR spectroscopy has been employed as a quality control tool for rapid screening and selection of flavor uniformity in the commercial production of black pepper oil [91]. IR spectra of the major black pepper oils are given in this chapter. Also, compounds responsible for the color of black "ripe" olives were identified by IR spectros- copy. The IR spectral data indicated that oleuropein was the com- pound responsible for the color problem [92].

Pyne and Wick [93] isolated the volatile components of tomatoes by distillation and identified them by IR spectroscopy as two un- saturated carbonyl compounds, trans-2-hexene and an aldehyde. Cis- 3-hexenol was responsible for the strong tomato aroma. Similarly, peach flavor was studied by Jennings and Sevenants [94]. They isolated the major components of peach essence and identified them by matching their IR spectra with known compounds. Volatile compounds such as benzaldehyde, benzyl alcohol, γ-caprolactone, γ-octalactone, γ-deca-lactone, and δ-decalactone were identified as the components of peach essence. The author has also studied pear flavor. IR spectra indicated that many esters in pears act as "character impact compounds." The methyl ester of trans-2, cis-4 decadienoic acid and hexyl acetate contribute to flavor [95].

The major volatiles in apple juice were examined by MacGregor et al. [96]. It was hoped that these data would provide a basis for the definition and differentiation of desirable and undesirable flavors in apple wine. Thirty compounds, tentatively identified on the basis of their IR spectra, included four aldehydes, one ketone, eleven alcohols, ten esters, and four fatty acids.

VII. ANALYSIS OF VITAMINS

The use of IR spectroscopy in the analysis of vitamins has been limited because of their ease of analysis by bioassay, u.v., and visible spectrophotometry. Morris and Haenni [97] have reported a rapid and sensitive IR technique for differentiating vitamins D_2 and D_3. This unique method is based on the examination of the spectrum between 10 and 11 μm. There is a strong absorption band at 10.3 μm (970 cm^{-1}) associated with the C-H bending vibration in the unsaturated side chain. The saturated side chain of vitamin D_3 does not exhibit this strong band. Figure 10 illustrates detailed spectra of vitamins D_2 and D_3. It is clear that the spectrum of vitamin D_3 exhibits a singlet at 10.4 μm while vitamin D_2 shows its strongest band at 10.3 μm with only a shoulder at 10.4 μm. This method can be made quantitative by use of a spectrophotometric difference technique. Recently, Parker [98] has reported IR spectra of many vitamins. His reports included Raman data as well as IR data; there is excellent agreement between them.

Recent evidence has pointed out the need for vitamin E in human diet. There are seven different forms of vitamin E with very similar

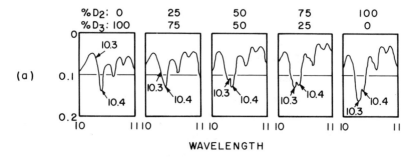

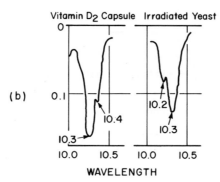

FIG. 10. (a) IR spectra, between 10.0 and 11.0 μm, of pure vitamin D_2, pure vitamin D_3, and mixtures of vitamins D_2 and D_3. (b) IR spectra, between 10.0 and 10.5 μm, of samples in which the form of vitamin D present can be determined by visual examination of the spectrum. (Reprinted from Ref. 99.)

TABLE 4

Distinguished Peaks in the IR Spectra of the Tocopherols

Form of tocopherol	Type of peak		
	Singlet (μm)	Doublet (μm)	Triplet (μm)
d-alpha	7.92 8.25	8.55, 8.65	8.97, 9.20, 9.40
d-beta	8.10	-	8.47, 8.55, 8.65[a]
d-gamma	8.65	9.07, 9.25 (8.05, 8.20)	-
d-delta	8.47 8.75 9.55	(8.05, 8.20)	-

[a]The components at 8.47 μm and 8.55 μm are shoulders. (Reprinted from Ref. [97].)

structures but their biological activities are significantly different. Alpha-tocopheral (vitamin E) is by far the most active form.
Until recently, there was no rapid analytical method for the separation and identification of vitamin E. Morris et al [99] took advantage of the slight differences in vitamin E structure and applied
IR spectroscopy for their analysis. Quantitative determination of
the α form of vitamin E in many vegetable oils such as soybean and
corn oil was accomplished by using their IR spectra. Some useful IR
bands for different forms of vitamin E are shown in Table 4.

VIII. ANALYSIS OF ADDITIVES

From the legal and regulatory standpoint, there are basically two
types of food additives -- those introduced into food during the
production and marketing of a crop (incidental additives), and those
utilized in converting raw agricultural commodities into processed
foods or "transformed food products" (intentional additives)[100].

Analytical methods for food additives are endless due to the
large number of additives utilized in food distribution channels.
Recently, the application of IR spectroscopy has become more widespread in the field of food additives; only a few examples will be
given here.

Silicones have been used in food as processing aids (antifoaming
agents). Due to their inertness and low concentration, silicones
have been difficult to identify using conventional analytical techniques. IR spectroscopy has been employed not only to determine the
type of silicone present but also its concentration. Figure 11 shows
the IR spectra of a silicone compound extracted from pineapple juice
[101]. IR techniques have also been applied to the analysis of silicone materials in bread, wafers, hydrolyzed vegetable protein, and
canned and frozen vegetables.

Weinkauff and co-workers [102] have developed an IR method for
the identification and differentiation of natural and synthetic caffeine. Specific IR absorption bands between 7 and 15 μm have been
assigned to natural and synethetic caffeine, and their differentia-
tions are shown in Table 5.

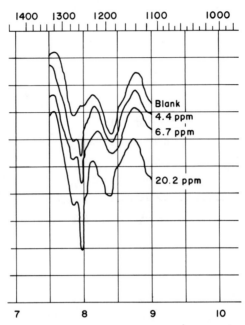

FIG. 11.   IR spectra of extracts from pineapple juice with and without silicone.   (Reprinted from Ref. 100.)

TABLE 5

Specific Absorption Bands for Differentiating Tea and Cocoa Caffeines from Urea Caffeines

Tea and cocoa sources (μm)	Synthetic source from urea (μm)
14.72 (Theobromine at 14.68)	14.51
14.23 Weak, broad	13.70 Strong
14.03	13.51 Strong
13.45	13.36
13.19	12.85 Weak
11.65 Broad region	11.42[a]
10.83	10.68[a]
10.45 Broad	10.54[a] (Strong doublet)
10.30 Broad region	9.77[a]
9.97	9.03
8.98	8.45
8.13 Strong	8.26 Strong
7.36 (Theobromine at 7.33)	7.91
	7.76
	7.48 Strong

[a]Prominent compound X bands. The table does not include any theobromine or theophylline bands. Bands listed for tea and cocoa are due to components that are extracted from tea and cocoa along with the theobromine. The 13.70- and 7.76-μm bands listed under synthetic are coincident with theobromine bands, but are not theobromine. (Reprinted from Ref. [102].)

Recently Seher [103] has published an excellent article on the analysis of nonionic surface-active agents using IR spectroscopy to identify the characteristic part of the molecule. His method should be applicable to identification of many emulsifiers used in food processing.

IX. CONCLUSION

It is clear from this brief review of the literature that the use of IR spectroscopy as an analytical tool in the area of food chemistry is important. Recent advances in sampling techniques, microanalytical separation and isolation, and instrumentation will combine to make IR spectroscopy one of the preferred qualitative and quantitative tools in the food industry. This should also be true for Raman spectroscopy.

The following references would be valuable for anyone involved in research and development in the food industry [19,42,104-108]. Table 6 is a summary of the most useful IR absorption bands for the major food components, i.e., water, fat, protein, and carbohydrates [104].

TABLE 6

IR Absorption Bands for Major Food Components[a]

Water:	1640 cm^{-1} (-OH deformation or bending)
	900-600 cm^{-1} (OH wagging causing tailing off)
Fats:	1750-1730 cm^{-1} (carbonyl, C=O stretching)
	1460 cm^{-1} (-CH$_2$ bending)
	1380 cm^{-1} (-OH bending)
	1230, 1150, 1100 cm^{-1} (triple peak due to C-O stretching)
	960 cm^{-1} (transunsaturation)
	720 cm^{-1} (-CH$_2$ rocking due to long chain hydrocarbons)
Proteins:	1660-1630 cm^{-1} (Amide I, C=O stretching)
	1570-1530 cm^{-1} (Amide II, N-H bending and C-N stretching)
	1450 cm^{-1} and 1350 cm^{-1} (CH$_2$ and CH$_3$ deformations)
	1300-1240 cm^{-1} (Amide III, C-N stretching and N-H inplane bending)
	720-600 cm^{-1} (Amide V, N-H out of plane bending; Amide IV and VI, skeletal vibrations)
Carbohydrate:	1875-1475 cm^{-1} (a number of double bond stretching deformations)
	1150-950 cm^{-1} (C-C and C-O vibrations)
	920 cm^{-1} and 770 cm^{-1} (ring stretching)

[a]Reprinted from Ref. 104.

REFERENCES

Moisture

1. C. O. Willits, *Anal. Chem.*, *23*, 1058 (1951).

2. W. R. Fetzer, *Anal. Chem.*, *23*, 1062 (1951).

3. H. E. Wistreich, J. E. Thompson, and E. Karmas, *Anal. Chem.*, *32*, 1054 (1960).

4. J. Mitchell, Jr., *Anal. Chem.*, *23*, 1069 (1951).

5. H. Brandenberger and H. Bader, *Anal. Chem.*, *33*, 1947 (1961).

6. H. J. Gold, *Food Technol.*, *18*, 586 (1964).

7. J. R. Hart, K. H. Norris, and C. Golumbic, *Cereal Chem.*, *39*, 94 (1962).

8. B. R. Rader, *J. Assoc. Offic. Agri. Chemists*, *49*, 726 (1966).

9. *Official Methods of Analysis of the Association of Official Agricultural Chemists* (11th ed.), Association of Official Agricultural Chemists, Washington, D. C., 1971.

10. J. D. S. Goulden and D. J. Manning, *J. Dairy Res.*, *37*, 107 (1970).

11. I. Ben-Gera and K. H. Norris, *J. Food Sci.*, *33*, 64 (1968).

12. J. Eisenbrand and K. Baumann, *Deutsche Lebensmittel Rundschau*, *6*, 169 (1969).

Carbohydrates

13. L. P. Kuhn, *Anal. Chem.*, *22*, 276 (1950).

14. R. L. Whistler and L. R. House, *Anal. Chem.*, *25*, 1463 (1953).

15. J. W. White, C. R. Eddy, L. Petty, and N. Hoban, *Anal. Chem.*, *30*, 508 (1958).

16. J. D. S. Goulden, *Nature*, 191, 905 (1961).

17. D. A. Biggs, *J. Assoc. Offic. Agri. Chemists*, *55*, 488 (1972).

18. M. Reintjis, D. D. Musco, and G. H. Joseph, *J. Food Sci.*, *27*, 441 (1962).

19. L. H. Meyer, *Food Chemistry*, Reinhold, New York, 1960.

20. W. W. Binkley, R. W. Binkley, and D. R. Diehl, *Intern. Sugar J.*, *73*, 783 (1971).

21. P. G. Kliman and M. J. Pallansch, *J. Dairy Sci.*, *50*, 1211 (1967).

22. E. Green, *J. Soc. Dairy Technol.*, *23*, 190 (1970).

23. J. M. Wilson, I. Ben-Gera, and A. Kramer, *J. Food Sci.*, *36*, 162 (1971).

24. R. W. Weik, *J. Assoc. Offic. Agri. Chemists, 55,* 257 (1972).

25. J. M. Wilson, A. Kramer, and I. Ben-Gera, *J. Food Sci., 38,* 14 (1973).

26. F. M. Lin and Y. Pomeranz, *J. Assoc. Offic. Agri. Chemists, 48,* 885 (1965).

27. H. Masakazur, *Bull. Chem. Soc. Japan, 43,* 3308 (1970).

28. A. J. Michell, *Australian J. Chem., 23,* 833 (1970).

29. J. E. Katon, J. T. Miller, and E. F. Bentley, *Carbohydrate Res., 10,* 505 (1969).

30. P. D. Vasko, J. Blackwell, and J. L. Koeing, *Carbohydrate Res., 19,* 297 (1971).

31. O. Horton, E. K. Just, and B. Gross, *Carbohydrate Res., 16,* 239 (1971).

32. J. E. Cluskey and Y. U. Wu, *Cereal Chem., 48,* 203 (1971).

33. H. P. Wehrli and Y. Pomeranz, *Cereal Chem., 47,* 160 (1970).

34. U. Buontempo, G. Careri, P. Fasella, and A. Ferraro, *Biopolymers, 10,* 2377 (1971).

35. R. C. Lord and N. Yu, *J. Mol. Biol. 50,* 509 (1970).

36. G. Careri, V. Mazzacurate, and G. Signorelli, *Physics Letters,* 31A, 425 (1970).

37. K. G. Brown, S. C. Erferth, E. W. Small, and W. L. Peticolas, *Proc. Nat. Acad. Sci. U.S., 69,* 1467 (1972).

Lipids

38. R. T. O'Connor, *J. Assoc. Offic. Agri. Chemists, 33,* 1 (1956).

39. O. D. Shreve, M. R. Heather, H. B. Knight, and D. Swern, *Anal. Chem., 22,* 1498 (1950).

40. J. B. Davenport, *Biochemistry and Methodology of Lipids,* Academic Press, New York, 1963.

41. N. K. Freeman, *J. Am. Oil Chemists Soc., 45,* 798 (1968).

42. D. Swern, *Bailey's Industrial Oil and Fat Products* (3rd ed.), John Wiley and Sons, Inc., New York, 1964.

43. F. E. Kurtz *Fundamentals of Dairy Chemistry* (B. H. Webb and A. H. Johnson, eds.), Avi Publishing, Westport, Connecticut, 1965.

44. R. G. Sinclair, A. F. McKay, G. S. Myers, and R. N. Jones, *J. Am. Chem. Soc., 74,* 2578 (1952).

45. R. P. Hansen, *J. Dairy Res., 36,* 77 (1969).

46. H. H. Perkins, Jr., *Textile Res. J., 41,* 559 (1971).

47. L. Biino, *Sci. Aliment., 15,* 58 (1969).

48. V. P. Fringeli, H. G. Muldner, H. Gunthard, W. Gasche, and W. Lenzinger, *Z. Naturforsch, B27,* 780 (1972).

49. J. H. Genrtz, *J. Intern. Chim. Anal., 16,* 216 (1967).

50. *Official Methods of Analysis of the Association of Official Agricultural Chemists* (11th ed.), Association of Official Agricultural Chemists, Washington, D. C., 1971.

51. *American Oil Chemists Society Tentative Method* (3rd ed.), Chicago, 1972.

52. A. Huang and D. Firestone, *J. Assoc. Offic. Agri. Chemists, 54,* 1288 (1971).

53. A. Huang and D. Firestone, *J. Assoc. Offic. Agri. Chemists, 55,* 47 (1971).

54. J. C. Bartlett and D. G. Chapman, *J. Agr. Food Chem., 39,* 50 (1961).

55. G. F. Bailey and R. J. Horvat, *J. Am. Oil Chem. Soc., 49,* 494 (1972).

56. R. G. Arnold and T. E. Hartung, *J. Food Sci., 36,* 166 (1971).

57. I. Ben-Gera and K. H. Norris, *Israel J. Arg. Res., 18,* 117 (1968).

Flavors

58. S. S. Chang in *Encyclopedia of Chem. Tech.* (2nd ed.), Vol. 9, John Wiley and Sons, Inc., New York, 1966, p. 336.

59. K. O. Herz and S. S. Chang, *J. Food Sci., 31,* 937 (1966).

60. R. A. Bernhard, *J. Food Sci., 26,* 401 (1961).

61. R. Teranishi, T. H. Schultz, W. H. McFadden, R. E. Lundin, and O. R. Black, *J. Food Sci., 28,* 541 (1963).

62. G. L. K. Hunter and W. B. Broyden, Jr., *J. Food Sci., 30,* 383 (1965).

63. G. L. K. Hunter and G. L. Parks, *J. Food Sci., 29,* 25 (1964).

64. G. L. K. Hunter and W. B. Broyden, Jr., *J. Food Sci., 30,* 1 (1965).

65. G. L. K. Hunter and W. B. Broyden, Jr., *Anal. Chem., 36,* 1122 (1964).

66. R. W. Holley, B. A. Stoyla, and D. Holley, *Food Res., 20,* 326 (1955).

67. N. T. Chu, F. M. Clydesdale, and F. J. Francis, *J. Food Sci., 38,* 1038 (1973).

68. T. Philip, *J. Food Sci., 38,* 1032 (1973).

69. R. M. Ikeda and E. M. Spitter, *J. Agr. Food, Chem., 12,* 114 (1964).

70. T. H. Schultz, R. Teranishi, W. H. McFadden, P. W. Kilpatrick, and J. Corse, *J. Food Sci., 29,* 790 (1964).

71. W. O. Macleod, Jr., and N. M. Buigues, *J. Food Sci., 29,* 565 (1964).

72. R. W. Wolford, G. E. Alberding, and J. A. Attaway, *J. Agr. and Food Chem., 10,* 297 (1962).

73. K. S. Rymal, R. W. Wolford, E. M. Ahmed, and R. A. Dennison, *Food Technol., 22,* 1592 (1968).

74. B. D. Moukherjee, R. E. Deck, and S. S. Chang, *J. Agr. and Food Chem., 13,* 131 (1965).

75. R. E. Deck and S. S. Chang, *Chem. and Ind., (1965), 1345.*

76. T. P. Dornseifer and T. J. Powers, *Food Technol., 19,* 877 (1965).

77. R. A. Scalan, R. C. Lindsay, L. M. Libby, and E. A. Day, *J. Dairy Sci., 51,* 1001 (1968).

78. B. W. Thorp and S. Patton, *J. Dairy Sci., 43,* 475 (1960).

79. G. A. Mank, J. Tobias, and R. M. Whitney, *J. Dairy Sci., 46,* 774 (1963).

80. S. Patton, *J. Dairy Sci., 37,* 446 (1954).

81. J. C. Underwood and V. J. Filipic, *J. Assoc. Offic. Agri. Chemists, 46,* 334 (1963).

82. I. Hornstein and P. F. Crowe, *J. Agr. and Food Chem., 11,* 147 (1963).

83. I. Hornstein and P. F. Crowe, *J. Agr. and Food Chem., 8,* 494 (1960).

84. R. J. Bouthilet, *Food Res., 16,* 137 (1951).

85. W. E. Kraulich and A. M. Pearson, *Food Res., 25,* 712 (1960).

86. O. F. Batzer, A. T. Santoro, M. C. Tan, W. A. Landmann, and B. S. Schweigert, *J. Agr. and Food Chem., 8,* 490 (1960).

87. C. H. T. Tonsbeck, E. B. Koenders, A. S. M. V. D. Zijden, and J. A. Losekoot, presented at the Flavor Symposium of the American Chemical Society, Atlantic City, September, 1968.

88. C. Hirai, K. O. Herz, B. R. Reddy, and S. S. Chang, presented at the National Meeting of Institute of Food Technologists, Philadelphia, May 19, 1968.

89. H. W. Ockerman, T. N. Blumer, and H. B. Graig, *J. Food Sci., 29,* 123 (1964).

90. D. A. Forss, E. A. Dunstone, E. H. Ramshaw, and W. Stark, *J. Food Sci., 27,* 90 (1962).

91. W. G. Jennings and R. E. Wrolstad, *J. Food Sci.*, *26*, 499 (1961).

92. K. L. Simpson, C. O. Chichester, and R. H. Vaughn, *J. Food Sci.*, *25*, 229 (1960).

93. A. W. Pyne and E. L. Wick, *J. Food Sci.*, *30*, 192 (1965).

94. W. G. Jennings and M. R. Sevenants, *J. Food Sci.*, *29*, 796 (1964).

95. W. G. Jennings and M. R. Sevenants, *J. Food Sci.*, *29*, 158 (1964).

96. D. R. MacGregor, H. Sugisawa, and J. S. Mathews, *J. Food Sci.*, *29*, 448 (1964).

Vitamins

97. W. W. Morris and E. O. Haenni, *J. Assoc. Offic. Agri. Chemists*, *45*, 92 (1962).

98. F. S. Parker, *Appl. Spectr.*, *15*, 96 (1961).

99. W. W. Morris, J. B. Wilkie, S. W. Jones, and L. Friedman, *Anal. Chem.*, *34*, 381 (1962).

Additives

100. H. J. Horner, J. E. Weiler, and N. C. Angelotti in *Instrumental Methods for Analysis of Food Additives* (N. H. Butz and H. J. Noebels, eds.), Wiley (Interscience), New York, 1961, p. 159.

101. D. F. McCaulley and J. W. Cook in *Instrumental Methods for Analysis of Food Additives* (N. H. Butz and H. J. Noebels, eds.), Wiley (Interscience), New York, 1961, p. 137.

102. O. J. Weinkauff, R. W. Radue, R. E. Keller, and H. R. Crane, *J. Agr. Food Chem.*, *9*, 397 (1961).

103. A. Seher, *Fette Seifen Anstrichmittel*, *71*, 138 (1969).

104. J. M. Wilson, Quantitative determination of the Composition of soybean samples, Ph.D. Thesis, Univ. of Maryland, 1971.

105. H. W. Schultz, E. A. Day, and L. M. Libby, *The Chemistry and Physiology of Flavors*, Avi Publishing Co., Westport, Connecticut, 1967.

106. B. H. Webb and A. H. Johnson, *Fundamentals of Dairy Chemistry*, Avi Publishing Co., Westport, Connecticut, 1965.

107. J. B. S. Braverman, *Introduction to the Biochemistry of Foods*, Elsevier, New York, 1963.

108. R. F. Gould, *Flavor Chemistry*, American Chemical Society, Washington, D. C., 1966.

Chapter 9

PETROLEUM

P. B. Tooke

Research Center
British Petroleum Company
Sunbury-on-Thames
Middlesex, England

I.   INTRODUCTION. 667

II.   SCOPE OF INFRARED APPLICATIONS. 669

III. FUELS 670
      A.   Motor Gasolines 670
      B.   Other Fuels 672

IV.   LUBRICANTS. 672
      A.   Mineral Oil Lubricants. 672
      B.   Synthetic Oil Lubricants. 688
      C.   Water-based Lubricants. 691
      D.   Greases 691

V.   RAMAN SPECTROSCOPY. 697

VI.   FUTURE DEVELOPMENTS 699

      REFERENCES. 700

I.   INTRODUCTION

A casual reader of the current analytical literature could hardly be
blamed for concluding that infrared (IR) and Raman spectroscopy are
rarely used in petroleum analysis. The relatively small number of
papers on such applications does reflect that, compared to a period

of 20 years ago, IR spectroscopy at least has, after an initial phase
of exploitation, earned a respected position alongside other methods
which are now regarded as more or less routine.

As an analytical tool, IR made its first really useful contributions to the petroleum field about 30 years ago. This was when
the technique was emerging from the point-by-point plotting stage to
the now commonplace recording type of instrument. A prime example
of IR analysis during that period was its use to examine aviation
gasolines during World War II. Subsequent availability of rapid recording double-beam spectrometers resulted in a deluge of petroleum
applications, with particular emphasis on the identification and
quantitative analysis of individual components in the gasoline range.
Today the technique is applied to such diverse products as fuels,
lubricants, additives, and a variety of fringe materials; and as a
quality control tool, and for monitoring and identifying various
pollutants.

In contrast Raman spectroscopy has experienced a checkered
history as an analytical method. During the 1930s, when IR was more
of an academic curiosity than a practical technique, Raman spectroscopy found quite considerable application in petroleum analysis,
particularly for materials in the gasoline boiling range. A decade
or so later, the roles of IR and Raman spectroscopy were reversed.
Raman methods were largely neglected in favor of the rapid and convenient IR spectrometer. As is now well known, Raman spectroscopy
was given a new impetus during the 1960s with the introduction of
the laser source, and a renaissance was forecast for the technique.
This has clearly come about in general. However, in specific relation to petroleum analysis it is apparent that laser Raman has not
made so dramatic an impact.

This chapter will therefore deal mainly with applications of IR
to the petroleum industry: inevitably the author's own experiences
will intrude, but the reader will rightly assume that all petroleum
companies have similar problems of analysis. The scope of IR applications will be considered first: then the analysis of fuels and

lubricants will be discussed in some detail. The chapter will conclude with a brief survey of the applications of Raman spectroscopy
to petroleum products.

II. SCOPE OF INFRARED APPLICATIONS

The major areas of IR analysis lie in the identification and measurement of additives, together with hydrocarbon components and types,
in fuels and lubricants. The IR analyst will also be expected to
examine many miscellaneous materials, a large proportion of which
are only indirectly related to petroleum. For example, current research into the production of protein from petroleum has meant that
the analyst must familiarize himself with the IR characteristics of
phospholipids, amides, amino acids, glycerides, and so on. Insecticide formulation and testing is another fringe activity with a biological bias. Although chromatographic methods are widely used, IR
finds application in monitoring concentrate blending, and also in
identifying biocides, e.g., Gore et al. [1].

Pure compounds and chromatographic fractions are common types
of samples arising within a petroleum research organization. Identification or at least type characterization requires a reliable and
comprehensive spectral library. Commercial collections of spectra
are invaluable, but they are unlikely to cover in any great depth
some particular area of interest. The analyst must therefore build
up his own collection which is based mainly on materials collected
within his own organization.

Gas chromatographic fractions generally present problems with
sample size. A large volume of literature exists on the application
of IR to such fractions. Some methods have been based on the examination of vapors using "on-the-fly" or stopped flow techniques, or
condensing out the vapor fraction and examining it as a liquid film
in solution or on potassium bromide. All methods, however, reduce
down to one problem - that of minimizing the effective cross-section
of the sample in the spectrometer beam. A powerful means of attack

in these small-sample, low-energy situations is Fourier transform
(FT) spectroscopy, and it is certain that this technique will assume
considerable importance in the future.

Many examples of the application of IR to petroleum analysis
could be cited here. Often these methods do not find their way into
the literature. They might be simple, unspectacular tests in which
IR measurements have replaced lengthy, and perhaps less precise,
conventional tests. The full extent of IR applications within the
petroleum industry cannot be demonstrated adequately here without
seriously trespassing on the rest of this volume. For example,
analysis of polymers, chemicals, environmental samples, and studies
of the chemistry of catalytically active surfaces are areas of great
importance within the industry, and IR is playing an important role
in all of these.

III. FUELS

IR methods have not found so great an application in the analysis of
fuels as compared to lubricants. Before the advent of gas chromatography (GC), IR was used, with limited success, for characterizing
narrow cuts in the gasoline range. Present day applications are
confined largely to additive detection and measurement.

A. Motor Gasolines

Anti-icing additives, e.g., alcohols and glycol ethers, may be
measured by IR. Three main approaches are possible. First, part
of the sample is washed free of additive with water, and is used as
a reference in a difference method measurement of the C-O stretch
in the 1050- to 1250-cm^{-1} region [2]. This method is relatively
insensitive, and not very applicable to current gasolines in which
the additive level is low. The second method, due to Ritchie and
Kulawic [3], again uses a similar water-washing/difference technique,
except that the gasoline is diluted with carbon tetrachloride so
that the additive OH groups are largely free rather than H-bonded.
The OH bands near 3600 cm^{-1} are quite narrow, and only the most

unlucky combination of additives would result in excessive overlap. The third method is more complex and demanding, and has been described by Jenkins and Scruton [4]. It requires that the gasoline be extracted with 1% of its volume of water. The extract, containing the additives, is then examined by difference against water alone over the C-O stretching range. This must be one of the few cases where the IR analyst deliberately chooses to record an aqueous solution spectrum! The method has adequate sensitivity; for example, dipropylene glycol can be determined at the 250-ppm level with a precision of 50 ppm or better. The main disadvantages are that (1) the water extraction must be carried out within narrow temperature limits in order to preserve a constant distribution coefficient, and (2) the calibration is nonlinear, and mutual interferences (which can however be corrected for) may arise if more than one anti-icing additive is present.

Gasoline additive packages, e.g., carburetor detergents and ignition control additives, are usually amenable to the methods outlined for lubricant additives (see Sec. IV). The detection of such additives in finished gasolines is, however, a very much more difficult problem simply because of the low concentrations, typically 0.1%, which are used. The first step prior to IR analysis is to remove the bulk of the gasoline, say 95%, by evaporation or preferably distillation. The residue can then be separated by dialysis, solvent partitioning, and so on, as for lubricants. One further complicating factor is that the residue will contain highly aromatic heavy ends, oxidation products, gums, and marker dye. All of these materials will contribute to the IR spectrum.

Benzene is an inevitable component of gasolines, and it seems likely that limits on the benzene content will soon be mandatory in many countries. Benzene has an intense IR absorption at 675 cm^{-1} which is unique to this compound and to no other aromatics in the boiling range. Even in cracked products, there is little absorption at this frequency from olefins, so that a nearly ideal situation exists for determining benzene contents by IR. The method is rapid and precise, and a detection limit of 0.1% is easily achieved.

B. Other Fuels

IR finds little application for fuels other than motor gasolines. Aviation fuels may be examined for anti-icing additives using methods similar to those above. Pour point depressant additives have been detected down to about 200 ppm in fuel oils by the difference method, but only if the additive-free fuel, taken from the same batch, is available.

IV. LUBRICANTS

The days of lubricants consisting solely of a refined mineral oil fraction have long since passed. Modern lubricants present the analyst with an ever-increasing challenge as the number and sophistication of additives grows. A cursory glance at the patent literature confirms the move away from simple additive packages to chemically more complex mixtures, and to the growing application of wholly synthetic lubricants. The IR spectroscopist can therefore no longer simply rely on recording a spectrum and hoping to interpret it: he must develop separation and concentration techniques enabling him to identify what are frequently only small amounts of additives from the lubricant matrix.

Lubricant analysis is perhaps the largest single field of application for IR in the petroleum industry. The pay-off from such analyses results from applications such as preliminary screening of new additives and lubricants before deciding if lengthy and expensive engine testing is worthwhile, monitoring lubricant blending, monitoring oil condition during service, and so on.

A. Mineral Oil Lubricants

The analyst is frequently asked to determine the hydrocarbon type composition of mineral base stocks. The IR spectrum of a base oil will generally allow a qualitative description such as paraffinic, naphthenic, or aromatic. The current tendency, however, is to give at least a partially quantitative description in terms of the relative

proportions of carbon atoms in aromatic, naphthenic, and paraffinic
structures, abbreviated to C_A, C_N, and C_P, respectively. The IR
spectrum shows absorptions specifically associated with aromatics
and acyclic paraffins, but naphthenes (i.e., cycloparaffins) exhibit
no reliable characteristic features. However, since C_A + C_N + C_P =
100%, C_N is calculable if C_A and C_P are known. The first serious
attempt at an IR measurement of these quantities was due to Brandes
[5], who correlated the 1600-cm^{-1} absorption with C_A and the 720-cm^{-1}
absorption with C_P. The standards chosen were base oils having
known values of C_A, etc., obtained by the n-d-M analysis of van Nes
and van Westen [6]. Other workers have attempted to modify the
original Brandes equations in the quest for better precision, but
it must be recognized that these methods are after all only rapid
substitutes for the n-d-M method, which itself is in no way an absolute measurement. The refinement of Brandes' method currently in
use at Sunbury [7] correlates C_A with bands at 1600 and 810 cm^{-1},
and C_P with the 720-cm^{-1} absorption, except that allowance is made
for aromatic interferences at this last frequency by correlation
via the 1600-cm^{-1} band intensity.

The majority of present day mineral oil lubricants contain
additives in concentrations lying typically in the range 0.1 to 10%.
It is immediately apparent that not all additives present in an oil
will necessarily show up in the IR spectrum, either because the concentration is below the detection limit or because of obscuration
by the base oil itself. One way of minimizing this latter effect is
to use the difference technique, first demonstrated for oils by
Powell [2]. It involves utilizing both beams of a spectrometer so
as to record the difference spectrum between the lubricant and its
base oil. Figure 1 shows how effective this method can be. Ideally
the difference technique will yield a spectrum arising solely from
the additives, although of course the spectrometer will "black-out"
in regions of high absorption by the base oil. Difficulties will
arise when the base oil is not available, for example, when analyzing
a competitor's product. In this case intelligent selection of a likely
base oil may suffice, although a rather uneven background might result.

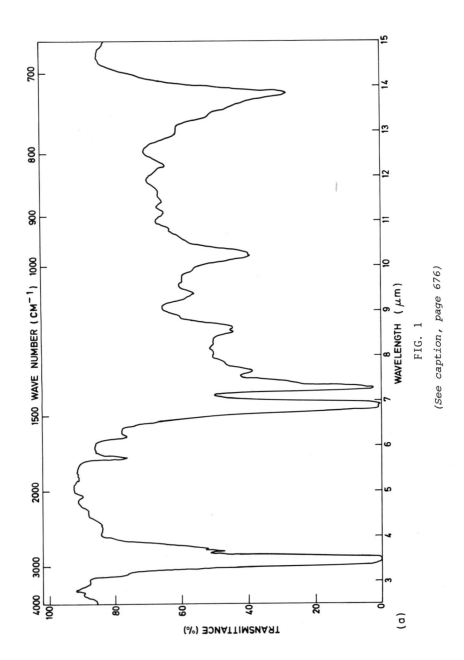

FIG. 1

(See caption, page 676)

(a)

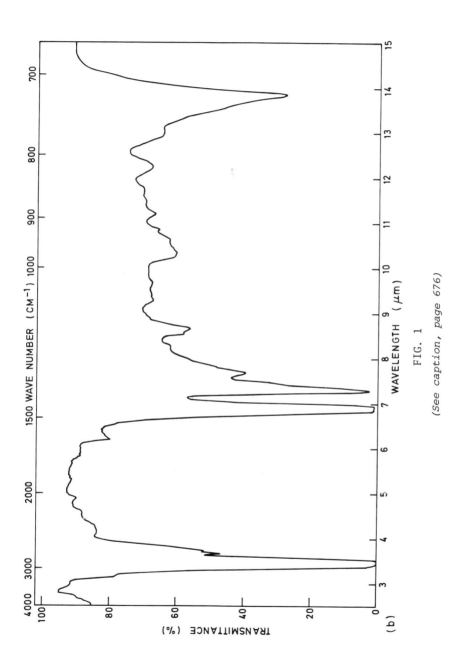

FIG. 1

(See caption, page 676)

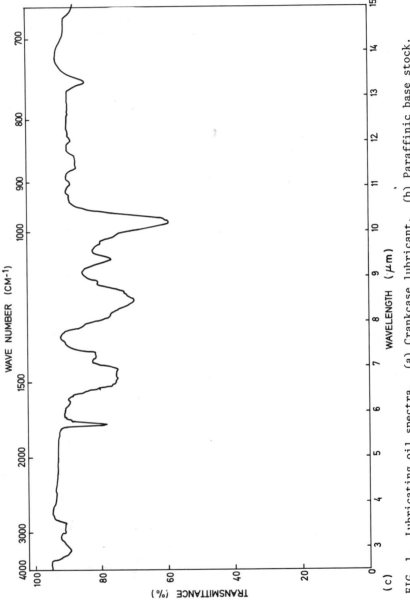

(c)

FIG. 1. Lubricating oil spectra. (a) Crankcase lubricant. (b) Paraffinic base stock.
(c) Difference spectrum between (a) and (b). Bands in (c) are due to additives.

Although the difference technique has the virtue of simplicity and will give a limited indication of composition, there is no real substitute for concentration, or preferably isolation, of the additives from the base oil. A variety of techniques may be applied to this end, but perhaps the most important is dialysis. The efficacy of this separation technique for petroleum additives has been described by Jenkins and Humphreys [8]. In essence, the oil is dialyzed through a rubber membrane (e.g., a balloon or finger-cot) held in a light solvent such as petroleum ether. Originally this was done by suspending the membrane in a large volume of solvent, but a quicker method which avoids the hazards of such an approach is to extract the membrane in a Soxhlet apparatus. Materials remaining inside the membrane, i.e., not passing into the solvent, are polymeric additives, those having an apparently high molecular weight as a result of micelle formation, and ionic compounds. Typical components found in such a dialysis concentrate would be polymethacrylates, polyisobutenes, polyamides, sulfonates, additive carbonates, metal carboxylates, and amine phosphates. An example is shown in Fig. 2(a) in which the oil of Fig. 1(a) has been dialyzed.

For many additives, dialysis gives a quantitative separation, an aspect which is useful when estimating additive concentration. Others present a less clear-cut situation; light ends of some polymers, notably polyisobutenes, will dialyze easily. Phenates and sulfurized glycerides behave somewhat unpredictably.

The dialysis concentrate is still a fairly complex mixture, and it is often necessary to break this down to aid further identification. Additive carbonates, almost inevitable components in multigrade oils, give strong, characteristic bands at about 1460 and 867 cm^{-1}, the former band often obscuring a significant part of the spectrum. Warming the concentrate with hydrochloric acid destroys the carbonate smoothly, and removal of the carbonate bands may reveal additional features in the spectrum; see Fig. 2(b). Acid hydrolysis will also aid or confirm identification of other components in the concentrate. For instance, soaps will yield carboxylic acids

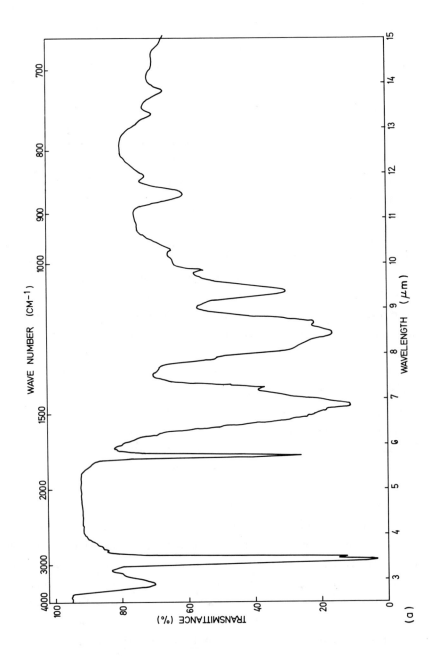

(a)

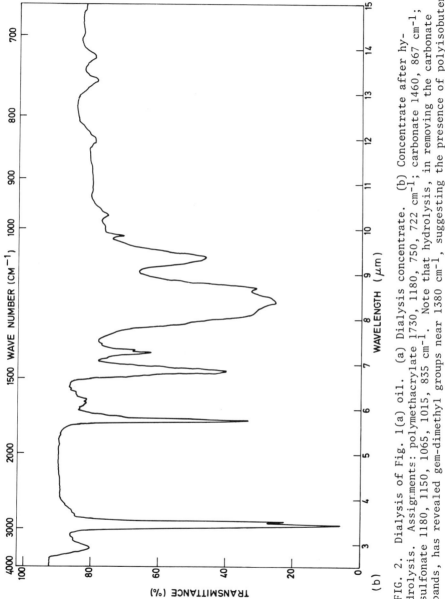

FIG. 2. Dialysis of Fig. 1(a) oil. (a) Dialysis concentrate. (b) Concentrate after hydrolysis. Assigrments: polymethacrylate 1730, 1180, 750, 722 cm⁻¹; carbonate 1460, 867 cm⁻¹; sulfonate 1180, 1150, 1065, 1015, 835 cm⁻¹. Note that hydrolysis, in removing the carbonate bands, has revealed gem-dimethyl groups near 1380 cm⁻¹, suggesting the presence of polyisobutene.

and amine salts will become amine hydrochlorides plus the parent
acid. Further dialysis of a hydrolyzed concentrate can give an additional separation and simplification of spectra.

It should be emphasized that during separations of this sort,
and indeed most of the other separations described below, IR spectra
are recorded at each stage. Not only does this keep the analyst in-
formed of his progress, but also indicates in which direction the
next stage of separation should proceed.

The dialyzate will contain, apart from the base oil, low-molec-
ular weight additives such as phenols, esters, amines, metal dialkyl
dithiophosphates, and aryl phosphates. A typical dialyzate is shown
in Fig. 3(a), derived from the oil of Fig. 1(a). The only obvious
feature in this example is the peak near 980 cm^{-1}, which can be
tentatively assigned to some type of phosphate. Further examination
of the dialyzate clearly requires further separation of the additives.

The simplest method for at least partially extracting polar
additives is to partition the dialyzate between two immiscible liquids
of very different polarities, for example, methanol and isooctane.
The bulk of the mineral oil will remain in the isooctane while polar
materials will preferentially migrate to the methanol phase. Figure
3(b) shows the effect of this separation; the spectrum of the methanol fraction now confirms a metal dialkyl dithiophosphate in the
dialyzate, and in addition has brought to light alkyl phenols.

Although this last method is simple and fast, additive complexity
may call for a more refined separation. A well-established method
is solid-liquid chromatography, in which the sample is eluted from
a stationary phase, e.g., silica gel, with a series of solvents of
increasing polarity. The degree of separation achieved depends very
much on the types of additives present and on the gradation of solvent polarities. A simplified method is to elute from silica gel
with pentane, benzene, and then diethyl ether; Fig. 4 shows three
such fractions obtained from the dialyzate of Fig. 3(a). An almost
complete resolution between the dithiophosphate and phenols is apparent. It should be noted that this type of chromatography is best
limited to dialyzates; attempts to separate the whole oil frequently

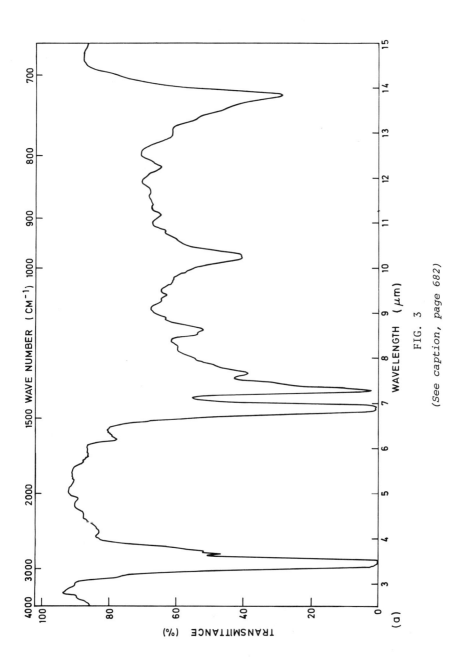

FIG. 3

(See caption, page 682)

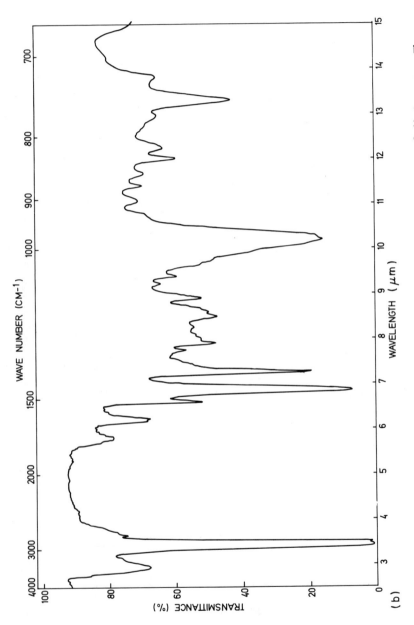

FIG. 3. Dialysis of Fig. 1(a) oil. (a) Dialyzate. (b) Methanol extract of dialyzate. The prominent band at 980 cm-1 is due to a dialkyldithiophosphate. Most of the other absorptions in (b) are associated with alkyl phenols.

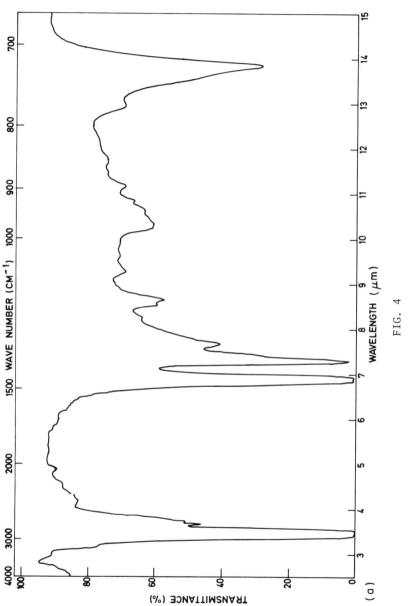

FIG. 4

(See caption, page 685)

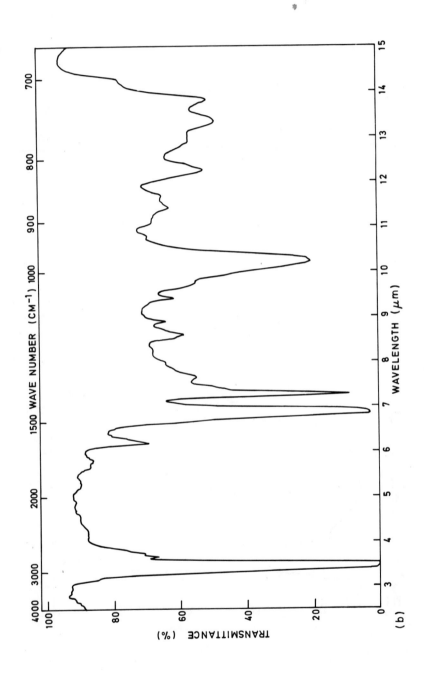

(b)

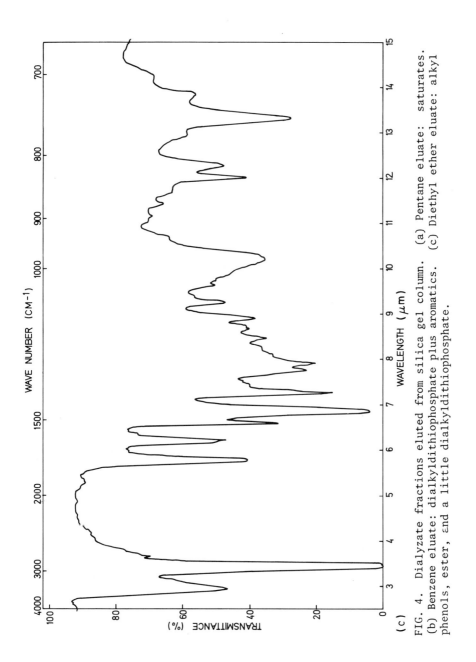

FIG. 4. Dialyzate fractions eluted from silica gel column. (a) Pentane eluate: saturates. (b) Benzene eluate: dialkyldithiophosphate plus aromatics. (c) Diethyl ether eluate: alkyl phenols, ester, and a little dialkyldithiophosphate.

can end in disaster, because high-molecular weight materials which
would otherwise have been removed by dialysis are not eluted in a
regular or predictable manner.

All these methods of separation can be applied equally well to
additive packages. The analysis is in this instance rather simpler
because the package may typically contain only 50% of a mineral oil,
if in fact any. Regardless of whether the sample is an oil or a
package, however, the analyst would be wise to recognize that IR will
not always give the final and complete answer on composition. It is
extremely useful to have available basic data such as the metals
which are present, or significant amounts of nitrogen, phosphorus,
sulfur, or chlorine. In addition, information as to the type of
oil, e.g., extreme-pressure, multigrade, heat transfer, etc., gives
the analyst some guidance on what sort of additives to expect.

Quantification of additives in lubricating oils is not so
straightforward as qualitative identification. Provided that a
reasonable match to the base oil is available and there is no interference with the analytical band by the base oil, then quantitative
measurements by the difference technique can be expected to have a
precision of about 5%. This is true for additives such as hindered
phenols, esters, and carboxylic acids. Quite often, though, the
additive to be quantified has an ill-defined composition which will
vary between batches. Problems will then arise in obtaining a rep-
resentative calibration. Under these circumstances poorer precision
must be expected.

Quality control of lubricants is a special case of quantitative
IR analysis - special because a particular and probably critical addi-
tive-blending specification must be monitored. In order to achieve
good precision, the analyst must be sure that he has available rep-
resentative samples of both the additive and the base oil. The effects of normal variations in these can be quite marked. For example,
measurement of the dispersant polyisobutene content of gas engine
oils at the 8% level, using the 1710-cm^{-1} band, can be in error by
as much as 10 to 25% (relative), unless the particular batches of
additives and base oil used are available for calibration purposes.
Also, it has been demonstrated by Tooke and Wilde [9] that the

aromaticity of the base oil can have a significant effect on measurements of hindered phenols in, for example, turbine oils.

Used lubricating oils can present difficulties in both qualitative and quantitative terms. The oil may contain substantial amounts
of soot and other solids (sometimes even metal fragments!). It will
probably have oxidized and even possibly cracked. Some or all of the
additives will have been depleted and degraded to different structures.
In addition, contamination with water, antifreeze, and unburnt fuel
may have occurred. Such samples are a daunting prospect, to say the
least. The initial step would be to remove volatile components such
as water, glycol, and fuel by evaporation and/or azeotroping. If the
sample contains so much solid matter that an IR spectrum cannot be
obtained, then filtration and/or centrifugation must be used. The
cleaned-up oil can then be qualitatively examined as usual, but it
must be borne in mind that some of the materials separated may be degradation products or possibly contaminants. For example, additive
carbonates might react with acidic oxidation products to form soaps
which would ultimately, and misleadingly, appear in the dialysis
concentrate.

Quantitative estimation of additives in used lubricants is
fraught with difficulties. For example, hindered phenols are normally
estimated using the OH band at 3650 cm^{-1}. Degradation of a hindered
phenol can proceed in several ways, one of which can result in a
modified phenolic structure having an OH band at 3640 cm^{-1}. The
simultaneous presence of both the original phenol and its degradation
products results in a band overlap situation which may completely
rule out quantitative estimation. Another example worth quoting is
that involving a common extreme-pressure additive based on a polymeric
sulphurized butene. Three possible bands are available for quantitative analysis, viz., 1120, 965, and 900 cm^{-1}. During use the additive
is depleted but, unfortunately, these three bands do not reduce all
in the same proportion. Which of the three possible answers should
the analyst then select?

An increasingly common application of IR analysis is in the
monitoring of oil condition during service. This simple measurement
is based on obtaining the difference spectrum of the oil vs. its

unused counterpart; the method has been described by Sarkis and
Schnack [10]. The spectrum obtained may show absorption associated
with carbonyl oxidation products, contamination with antifreeze, and
reaction of the oil with blow-by gases to form nitrates and nitro-
compounds. Associated intensities will generally show a regular
increase with service time. It is quite possible to correlate band
intensity with oil condition and to devise empirical absorbance
limits which, when reached, indicate that an oil change is due.
Being simple and fast, this is therefore an attractive test method,
but it must be applied with caution. For example, oils under different conditions of use may, at the end of their useful lives, have
very different absorbance limits, so that this IR method must be
applied specifically rather than universally.

B. Synthetic Oil Lubricants

The commonest synthetic lubricating oil base stocks are esters
and polyethers. Either type is easily recognizable from its IR
spectrum, giving very strong characteristic absorptions. Two typical
examples are shown in Fig. 5.

Additive identification in such oils is a difficult problem,
one to which IR methods cannot be easily applied. The main problem
is the intense absorption of the base oil together with the particular
additive types encountered. Typical additives are amines, amine salts,
other nitrogen-containing materials such as phenothiazines and benzo-
triazoles, and phosphates. Dialysis may give partial success, for
example, in concentrating amine salts from ester-type oils, but inevitably a small amount of the oil will remain in the concentrate.
Polyether-type oils dialyze only partially, if at all, and usually
little is to be gained here from this separation.

The difference technique can find some application to these
lubricants, but it is all-important that a very good matching base
oil is available. Additive detection sensitivity is at least an
order of magnitude worse than for mineral oils, due to the intense
background absorption and the consequent need to work at small pathlengths.

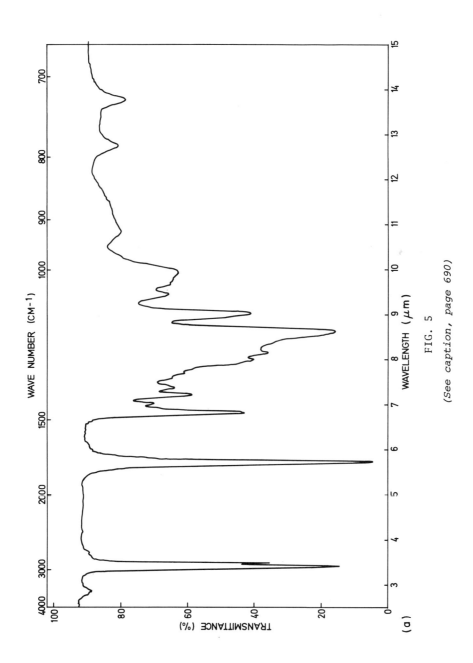

FIG. 5

(See caption, page 690)

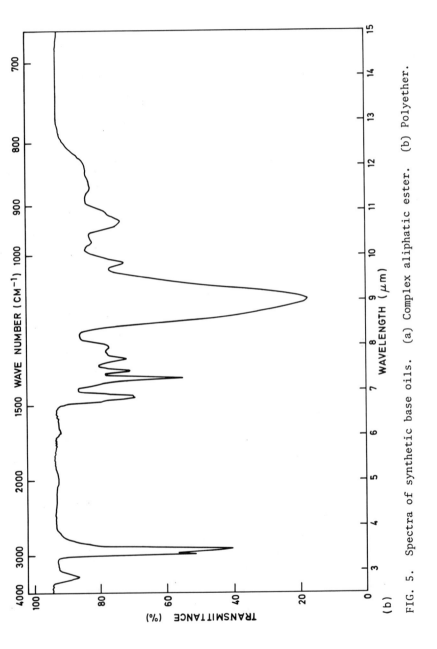

(b)

FIG. 5.   Spectra of synthetic base oils.   (a) Complex aliphatic ester.   (b) Polyether.

One method which is worth mentioning, and which can be applied
to both synthetic and mineral oils, is the use of ion exchange resins
for isolating amines. By percolating the oil through a resin in its
acid form, amines are retained and the resin can be washed free of
oil. Subsequent elution of the resin with a relatively strong base
such as ammonia yields the amine, which is then identified from its
IR spectrum.

C. Water-based Lubricants

Water-based lubricants vary enormously in composition. Apart
from water, they may contain such things as alkanolamines, sodium
nitrite, metal soaps and sulfonates, and mineral oil. Water is
traditionally anathema to the IR spectroscopist. Apart from its
disastrous effects on alkali halide cells, water has so strong an
absorption spectrum as to completely obscure most absorption bands
arising from other components in solution. The first step then is
to remove the water from the sample. This is readily achieved by
distillation, although there is always the risk that some components
may in part be carried over into the distillate. The water fraction
should therefore be acidified with hydrochloric acid and extracted
with a hydrocarbon solvent to recover nonbasic materials such as
odorants. The (acidified) water fraction can then be evaporated,
alkanolamines or amines recovered as their hydrochlorides, and subsequently characterized from their IR spectra [11].

The distillation residue can be extracted with acetone at 0°C.
Typically, alkanolamines are extracted to leave sodium nitrite and
soaps. These latter can be examined in more detail after destruction
of nitrite with ammonium sulfate and subsequent hydrolysis of soaps
to parent acids. Figure 6 shows some spectra of fractions obtained
from a water-based lubricant.

D. Greases

A grease is basically an intimate mixture of an oil and a thickener. The oil is generally naphthenic or paraffinic mineral oil,
although synthetic oils are becoming more common. The thickener can

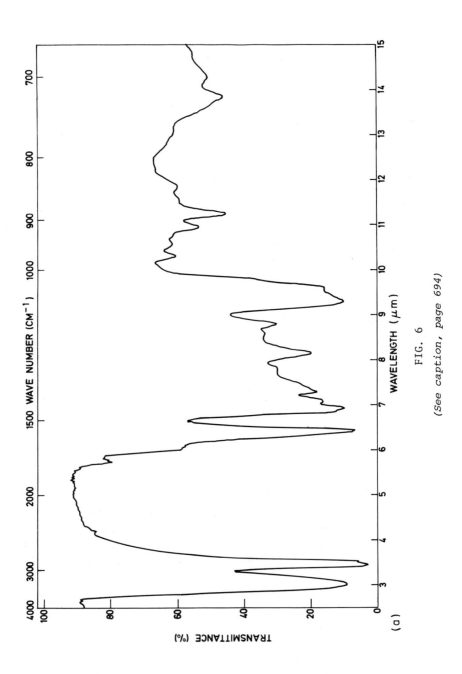

FIG. 6

(See caption, page 694)

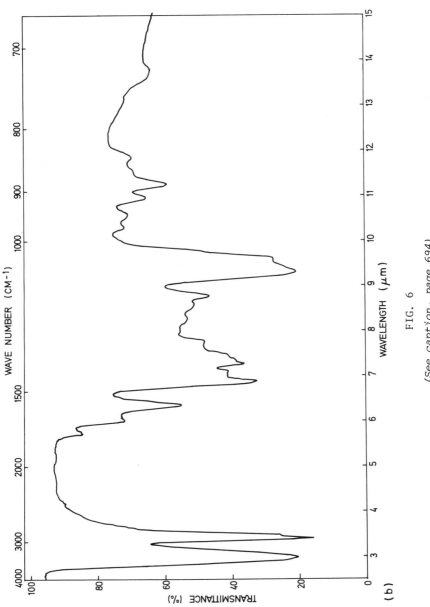

FIG. 6

(See caption, page 694)

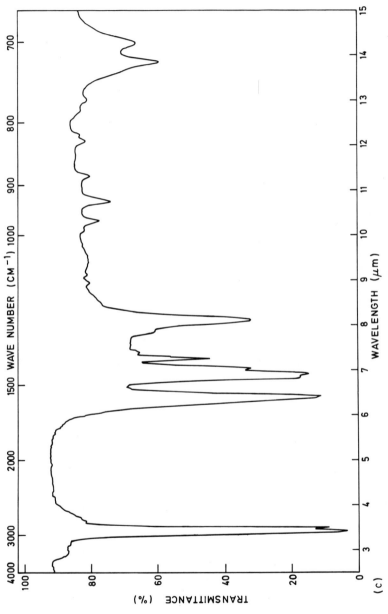

FIG. 6. Water-based lubricant spectra. (a) Residue after removing water. (b) Acetone extract
of residue: diethanolamine. (c) Insoluble material from extraction: metallic soap (1560 cm^{-1}) and
sodium nitrite (1230 and 829 cm^{-1}).

be a mineral such as silica or modified (organophilic) aluminosili-
cate, a metallic soap, a polymer, or graphite. In addition there
might be small amounts of additives such as amines, organic phosphates, esters, and sodium nitrite.

The IR spectrum of a grease will give a general idea of the
major components, as shown in Fig. 7, but in most cases separation
or concentration techniques are required. Dialysis has been used
very successfully for separating the oil and filler components, although a rather more lengthy extraction is required than is the
case for mineral oil lubricants. Dialysis cannot, however, be applied to greases made with polyether-type oils.

The dialysis concentrate will contain minerals, soaps, etc.,
which give broadly characteristic spectra. Mineral fillers may
sometimes be identified, e.g., silica and bentonites. Metallic
soaps show strong absorptions in the range 1520 to 1620 cm^{-1}, the
exact position being strongly dependent on the metal atom. Jenkins
illustrated the effect of the metal on band position [12], and a
later paper by Elliot and Harting [13] has confirmed that other factors such as crystalline form and particle size of the soap may have
perturbing effects on these bands. The most common soaps are stear-
ates and hydroxystearates. They are readily discriminated by their
spectra. The parent acid can in any case be easily regenerated
from the soap and characterized, if not identified, by IR. Alumin-
ium soaps are apparently unique in showing a strong absorption at
about 998 cm^{-1}.

Sodium nitrite will, if present, appear in the dialysis concentrate. It is identified from its bands at 1230 and 829 cm^{-1}.
Graphite-based greases are difficult to cope with; it will not prove
possible to obtain a spectrum from the concentrate because of scattering and absorption by the graphite. Soaps at least can be identified indirectly in these cases by hydrolysis, followed by extraction
and IR examination of the parent acid so formed.

The dialyzate will be largely mineral oil containing small
amounts of additives and is treated as for lubricating oil dialyzates.

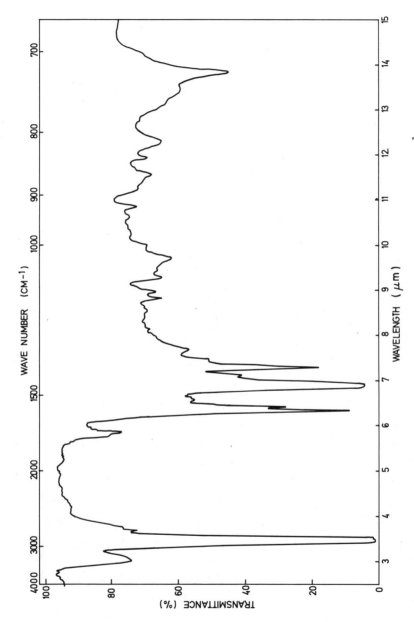

FIG. 7. Spectrum of a mixed-soap grease. Lithium soap, 1577 and 1555 cm^{-1}; lead soap, small band near 1525 cm^{-1}.

V. RAMAN SPECTROSCOPY

The Raman effect was first demonstrated in 1928. Despite considerable experimental difficulties, such as large sample size requirement, sample fluorescence, and long exposure times, the technique was exploited with some enthusiasm. This was hardly surprising, since Raman spectroscopy was virtually the only practical means available at that time for studying molecular structure. Much of the reported work was on the structure of hydrocarbons, particularly relevant to petroleum analysis, and up to about 1950 it was limited mainly to compounds and fractions boiling well below 200°C. A good example of that period was a paper by Heigl et al. [14], in which the authors described a method for the quantitative analysis of aromatics and olefins in naphthas. Accuracy was about 10% and carbon tetrachloride was used as an internal standard.

Later, the Raman technique was extended to higher-boiling fractions, but not without some considerable effort being given to pretreating the sample so as to minimize fluorescence. In 1953, Robert [15] succeeded in recording the spectra of gas and lubricating oils. Saturate fractions were shown to have differing Raman spectra, although their IR spectra were very similar. Robert attributed these differences to variations in naphthene types and concentrations.

During the 1950s, other methods of analysis began to gain importance, notably IR, mass spectrometry (MS), and GC. As a consequence, Raman spectroscopy lost ground as an analytical tool and was relegated to a largely academic role.

The resurgence of interest during the 1960s led to the appearance of many commercial laser Raman spectrometers. The available types now roughly parallel IR instrumentation, ranging from simplified bench-top spectrometers to highly sophisticated triple monochromator designs. The result has been that in general it is now a fairly straightforward matter to obtain a Raman spectrum from quite small quantities of material. Despite these considerations, however, laser Raman spectroscopy has received support mainly from academic establishments. There has been only limited activity so far in applying the technique to practical problems of analysis.

Based on the author's own experience, petroleum fractions show
the following general features in their Raman spectra:

1. Low-boiling fractions, e.g., gasoline range, show a wealth
of detail, although (as in the IR) the spectra tend to be dominated
by aromatic structures.

2. Detail is rapidly lost in moving up the boiling range.
Lubricating oils, for example, show broad, ill-defined bands with
very little "structure."

3. Fluorescence effects intrude with heavier products and may
completely mask the true spectrum even after careful sample pretreat-
ment and use of red light excitation.

These effects have led to the (personal) view that laser Raman
is not going to contribute very much to the analysis of petroleum
and its fractions. In this particular field the availability of
other techniques leaves Raman spectroscopy more as a method which
may on occasion contribute some circumstantial evidence rather than
as a source of primary data.

Raman spectroscopy does, however, have potential in the examination of pure compounds, simple mixtures, and regular structures, i.e.,
polymers. Many functional groups of high symmetry, e.g., C=C, C=S,
S-S, are strongly Raman-active, but only weakly active in the IR.
Where such groups are present, Raman could furnish structural information which is not otherwise easily obtainable. Generally, however,
the occurrence of these symmetric groups is relatively uncommon in
petroleum analysis.

As a complement to IR, Raman is useful in the identification of
single compounds, although Raman spectral libraries are, compared to
IR, very limited as yet. The nanoliter sensitivity possible with
laser Raman may find application in the identification of GC fractions, in combination with IR and perhaps MS.

The areas in which laser Raman will prove really useful to the
petroleum industry fall, strictly speaking, outside the scope of
this chapter. Polymer stereochemistry, remote sensing of atmospheric

pollution, and the study of catalyst surfaces are just three fields
in which laser Raman spectroscopy has already made an impact, and
further progress can be expected with confidence.

VI. FUTURE DEVELOPMENTS

From an IR point of view, petroleum analysis stands to gain little
from further instrumental developments offering better resolution
and photometric accuracy than is presently available from medium-
priced grating spectrometers. This is also to some extent true for
automation of instruments, although of course in very specialized
applications in which repetitive measurements are the rule, automatic
sample handling and absorbance/concentration read-out would be desirable. However, the new FT instruments operating in the "fingerprint" region will undoubtedly have a tremendous impact. With a
sensitivity increase of the order of 100 times that of conventional
dispersive systems, FT IR spectroscopy will greatly benefit work in
situations where sample sizes or available energy are severely limited.
These areas are the study of catalyst surfaces, the identification
of chromatographic fractions, and the identification of small amounts
of impurities. In addition, subtraction techniques may provide additional information not available with dispersive instruments. At present these instruments are very expensive, but hopefully competition
and technical advances may alter this situation.

Partly as a result of the availability of cheap, routine, bench-
top IR spectrometers, the wider use of IR, in the sense of partial
decentralization from a specifically spectroscopic laboratory, can
be anticipated. Already many organic chemistry laboratories have
an IR spectrometer as a standard piece of equipment; quality control
and routine testing laboratories will also increasingly follow this
trend in the future. Although by no means a new technique, IR is now
gaining more and more recognition as a powerful and rapid test method,
and is gradually finding its way into official methods.

Although a rather gloomy picture has been painted of the applications of Raman spectroscopy to the analysis of petroleum, it is
quite likely that this view will have to be revised in years to come.

Development of even cheaper, simplified routine instruments seems probable, and the wide subsequent use of these could result in the discovery of alternative or novel solutions to analytical problems. Also, the troublesome effects of sample fluorescence may eventually be overcome with the use of pulsed lasers and gated detector systems. Whatever happens, it is improbable that Raman spectroscopy will ever return to the limbo of its prelaser days.

ACKNOWLEDGMENT

Permission to publish has been given by The British Petroleum Company Limited.

REFERENCES

1. R. C. Gore, R. F. Hanna, S. C. Pattacini, and T. J. Porro, *J. Assoc. Offic. Anal. Chem., 54,* 1040 (1971).

2. H. Powell, *J. Appl. Chem., 6,* 488 (1956).

3. R. K. Ritchie and D. Kulawic, *Anal. Chem., 42,* 1080 (1970).

4. G. I. Jenkins and M. R. Scruton, *J. Inst. Petrol., 49,* 176 (1963).

5. G. Brandes, *Brenstoff-Chem., 37,* 263 (1956).

6. K. van Nes and H. A. van Westen in *Aspects of the Constitution of Mineral Oils,* Elsevier, Amsterdam, 1951.

7. P. B. Tooke, B. P. Research Centre Technical Memorandum No 110 194, available National Lending Library, Boston Spa, Yorkshire, England.

8. G. I. Jenkins and C. M. A. Humphreys, *J. Inst. Petrol., 51,* 1 (1965).

9. P. B. Tooke and A. J. Wilde, *J. Inst. Petrol., 55,* 343 (1969).

10. A. B. Sarkis and D. D. Schnack, Optimum Gas-Engine and Lubricating-Oil Performance Through Modern Analytical Techniques, Paper No. 64-OGP-16 of ASME Oil and Gas Power Conference and Exhibit, Dallas, Texas, April 1964.

11. C. S. Douse and P. B. Tooke, *Can. J. Spectr., 18,* 101 (1973).

12. G. I. Jenkins, Grease Analysis using Infra Red and Ultra Violet Spectrometry, Paper 709B of SAE International Summer Meeting, Montreal, June 1963.

13. J. J. Elliot and G. L. Harting, *NLGI Spokes, 34,* 84 (1970).

14. J. J. Heigl, J. F. Black, and B. F. Dubenbostel, *Anal. Chem., 21,* 554 (1949).

15. L. Robert, *Spectrochim Acta, 6,* 115 (1953).

Chapter 10

TEXTILES

Gultekin Celikiz

Department of Chemistry
Philadelphia College of Textiles and Science
Philadelphia, Pennsylvania

I.   INTRODUCTION. 702

II. METHODS FOR OBTAINING INFRARED SPECTRA. 702
     A. Films from Solution 702
     B. Melts 703
     C. Mineral Oil Mulls 703
     D. Alkali Halide Pressed Disks 705
     E. Frustrated Multiple Internal Reflectance. 707
     F. Infrared Microtechniques. 708
     G. Thin Fiber Films. 708

III. APPLICATIONS OF INFRARED SPECTROSCOPY 709
     A. Identification of Fibers. 709
     B. Identification of Finishes, Plasticizers,
         Additives, and Coatings 709
     C. Studies of Chemical Modification of Cellulose . . . 710
     D. Chemical Modification of Other Textiles 711
     E. The Interpretation of Textile Spectra 711
     F. Identification of Textile Dyes. 712

IV. CONCLUSION. 712

     REFERENCES. 713

I. INTRODUCTION

Although this book considers both infrared (IR) and Raman spectroscopy, the present chapter on textiles will be limited to IR spectroscopy only. Raman spectroscopy has not made any significant inroads
into the area of textile analysis as yet. Laser Raman spectroscopy
is useful in many areas, including polymer chemistry, and this is
discussed in Chap. 12.

IR spectroscopy has been a powerful analytical technique for
chemists for many years. However, it was not until the 1950s that
textile chemists began to utilize this method of analysis in the
laboratory. The lag in applying IR spectroscopy to textile analysis
resulted from the difficulty in using conventional techniques with
textile materials. New sampling techniques were introduced between
1950 and 1973 which have made IR spectroscopy a common analytical as
well as research tool for textile chemists. I will first review the
different methods of sample preparation and then discuss some applications to textile analysis.

II. METHODS FOR OBTAINING INFRARED SPECTRA

A. Films from Solution

One of the first textile applications of IR spectroscopy was
reported by Rowen et al. [1]. They were able to cast film of cellulose acetate from acetone. Because of the nonuniformity of the film
thickness, percent transmission of various bands varied from one
area of the film to another area of the same film.

Any film-forming solid which is soluble in a volatile solvent
can be prepared for analysis by casting film from solution and evap-
orating off the solvent. These films can be cast directly on an
alkali halide plate if the material is solvent-soluble or on a silver
chloride plate for water-soluble materials. Irtran-2 plates are
also available for this purpose but they are more expensive.

Low-boiling solvents are best, but if a high-boiling solvent is
used it can be removed by vacuum evaporation. This method is not
widely used since it is laborious and time-consuming.

B. Melts

If the sample under investigation has a low melting point, it is possible to prepare a melt of the sample between salt plates. Most low-melting samples will give good spectra. The bands will be sharp and well defined. However, since the crystallization may take place with a preferred orientation, the resulting spectra may be different from ones obtained by using other methods.

C. Mineral Oil Mulls

One of the oldest methods of obtaining an IR spectrum of a solid is the mineral oil mull technique. It is also known as the Nujol mull method because of the trademark of a well-known brand of puri-fied mineral oil. For most organic samples, a mineral oil mull is a suitable support. The only requirement is that the sample be in powder form.

In 1950, Forziati and co-workers at the Textile Section of the National Bureau of Standards developed a method of powdering cotton for IR transmission work [2]. After trying various forms of grind-ing, they decided to use a vibratory ball mill. After an 8-hr ex-traction in a Soxhlet with ethanol, the cotton sample was washed with distilled water and dried in air at room temperature. The dried sample was then ground in a Wiley mill to pass a 20-mesh screen. Of the ground cotton, 5 g was dried for 1 week over magnesium per-chlorate and finely ground in a modified vibratory mill. The cotton, thus powderized, was mulled in mineral oil and the resulting paste was smeared on a demountable rock salt cell for IR absorption measure-ment. Forziati and his co-workers arrived at the following conclu-sions:

1. The disintegration of fibers was rapid during the first 15 min. Microscopic examination showed no further reduction in size after 30 min of milling.

2. Milling resulted in practically no oxidation of cellulose.

3. There is a marked decrease in the degree of polymerization.

4. Crystalline regions of cellulose were converted to amorphous regions as a result of grinding.

According to the workers from the National Bureau of Standards, there were no significant differences in the IR spectra of cotton ground in a Wiley mill and passed through a 200-mesh screen and cotton ground in a vibratory ball mill for an 8-hr period.

Preparation of a satisfactory mull is both an art and a science. One cannot master the technique overnight. Some investigators suggest the use of an agate mortar and pestle, but this author has found that a combination of an inexpensive coors spot plate and a 5-in. test tube will usually work quite well with ground cellulosics. A few milligrams of the powdered sample is placed on the plate. A drop of mineral oil is then added to the center of the sample. Using the outside of the test tube like a pestle, the sample is ground and dispersed in the mineral oil. Depending on the consistency, more oil or more sample can be added. During mulling, the mixture is usually scraped from both the plate and the end of the test tube with a clean spatula. It is then collected in the center of the plate and the mixing operation is repeated until a finely dispersed paste is obtained. A small quantity of this paste is transferred to one of the sodium chloride plates of a demountable cell, and a sandwich is made with a second sodium chloride plate. The plates are placed in a suitable cell holder and the sample is scanned. If the mull produces a spectrum that is too intense, the sodium chloride plates are separated and, with the aid of a soft facial tissue, some of the sample is wiped gently from the surface of one of the plates. The two sodium chloride plates are sandwiched again and the sample is scanned. On the other hand, if the prepared mull produces too weak a spectrum, the only solution is to add more sample to the spot plate and repeat the mulling. This author finds it helpful to grind the sample mechanically first in a vibrating mill such as the "Wig-L-Bug" before dispersing it in the mineral oil.

Some workers find it useful to add a volatile solvent as an aid in grinding [3]. The solvent is added to the sample and the sample is ground as the solvent vaporizes. Another useful technique in grinding amorphous polymers is to chill the sample in dry ice [3]. Kendall reports the use of acetone-dry ice mixtures or liquid nitrogen to chill the sample to help in grinding and prior to mulling [4].

This author has also used a small laboratory file to obtain filings from plastic samples as a preliminary step before mulling. The procedure has produced very good spectra.

There are some drawbacks to the mulling technique. Unless the sample is ground sufficiently, it will scatter IR radiation at lower wavelengths. Also, mineral oil itself has four absorption bands between 5000 and 650 cm^{-1} (2 to 15 μm). These are at 3000 to 2850 cm^{-1} (3.33 to 3.51 μm) due to C-H stretching vibrations, at 1468 cm^{-1} (6.81 μm) and 1379 cm^{-1} (7.25 μm) due to C-H bending modes, and at 720 cm^{-1} (13.88 μm), a weak band, due to methylene rocking vibrations. If the sample has absorption bands in these regions, the analyst should consider using a halogenated hydrocarbon as the mulling agent, primarily in the region from 3000 to 1350 cm^{-1}. Perfluorokerosene (Fluorolube) or hexachlorobutadiene are alternate materials for making mulls. Fluorolube has clear regions between 2 to 8 μm and 11 to 15 μm which can be used for qualitative analysis.

Another drawback of the mulling technique is that it is a qualitative method only. For textile fibers, there are other methods that are more satisfactory, and they will be discussed later.

D. Alkali Halide Pressed Disks

This procedure for sample preparation was introduced simultaneously by Schiedt and Remwein [5] in Germany and by Sister Stimson and O'Donnell [6] in the United States.

Harvey and co-workers were attempting to obtain IR spectra of relatively large solid particles of polyvinylchloride (PVC) [7]. Other sampling techniques were not very satisfactory. They managed

to run an IR spectrum of PVC by using the alkali halide pressed disk
method. Among the different alkali halide salts that they used,
potassium bromide gave the best spectrum with PVC. This is attributed to the similarity in the indices of refraction of potassium
bromide and PVC. When the index of refraction of the suspending
medium and that of the sample are matched, particle size reflections
are minimized. Harvey and co-workers also obtained IR spectra of
nylon and paper fibers in potassium bromide.

O'Connor et al. successfully applied the potassium bromide disk
technique in obtaining the IR spectra of cotton fibers, yarns, and
fabrics [8]. A sample of cotton or rayon is cut in a Wiley mill for
passing through a 20-mesh screen. A 1- to 2-mg sample is mixed with
350 mg of potassium bromide in a "Wig-L-Bug" vibrating mixer for 15
to 20 sec. A portion of the mixture weighing 300 mg is transferred
to a die, which is then assembled and evacuated to remove the trapped
air. The die is subjected to a total load of 20,000 lb force while
under evacuation for 1 min. If the potassium bromide is anhydrous,
it will yield an optically transparent disk about 1 mm in thickness.

Use of alkali halide salts other than potassium bromide to obtain an IR spectrum of cotton often requires excessive grinding of
the sample. Forziati and Rowen [9] in 1951 pointed out the changes
in the IR spectrum of cellulose because of changes in the crystallinity as a result of excessive grinding. O'Connor and co-workers [8]
showed that it is not necessary to grind cotton to pass through a
250-mesh screen in order to obtain good IR spectra. The principle
reason for this is the similarity between the indices of refraction
of cellulose and potassium bromide.

The potassium bromide disk method as reported by O'Connor and
co-workers is widely used by textile chemists to obtain IR spectra
of synthetic fibers as well as those of cotton and modified cotton.
It is important to remember that, unless the indices of refraction
of the fiber and potassium bromide are quite similar, one has to
grind the fiber very finely.

E. Frustrated Multiple Internal Reflectance

In 1959, Fahrenfort [10] introduced the attenuated total reflectance (ATR) technique. When IR radiation is transmitted through a crystal by total internal reflection, no spectrum is produced. However, an ATR spectrum does result when a sample with a lower index of refraction than that of the crystal is placed in intimate contact with the crystal, and light is transmitted through that crystal at an angle above the critical angle. The critical angle θ is the least angle of incidence at which total reflectance takes place. It may be calculated from the expression:

$$\sin \theta = \frac{n_s}{n_c}$$

where n_s = index of refraction of the sample

n_c = index of refraction of the crystal

One of the first crystals used successfully was thallous bromide-iodide (KRS-5) with an index of refraction, n_c, of 2.38 and an angle of incidence of about 40° with a sample having an index of refraction n_s of 1.55. The use of three other crystals has been developed. They are silver chloride with an index of refraction of 2, Irtran-2 with an index of refraction of 2.25, and germanium with an index of refraction of 4.02.

IR radiation enters the crystal and strikes the sample at an angle that is greater than the critical angle at the interface. It then returns to the crystal and passes through to the monochromator of the instrument. The ATR technique does not give strong spectra because it involves only a single reflection from the sample.

Frustrated multiple internal reflectance (FMIR) spectroscopy is an improvement on the ATR method. In the FMIR technique, IR radiation becomes trapped in the crystal and undergoes multiple internal reflectances from the sample. The effective depth of penetration into the sample ranges from 0.1 to 10 μm depending on the crystal element, the angle of incidence of the beam on the interface, and the wavelength of light.

The FMIR technique yields a spectrum that closely resembles the
IR absorption spectrum. However, as the wavelength is increased,
the bands in the spectrum tend to appear deeper than the corresponding bands in the absorption spectrum. Also, there is a very small
shift in the position of the maxima for FMIR. The differences are
minor and do not prevent textile chemists from using the FMIR technique in obtaining IR spectra and assigning bands by comparing them
with standard adsorption spectra. In order to obtain good spectra,
it is important that there should be very good contact between the
sample surface and the crystal [11]. More discussion on this technique will be presented in Chap. 11.

F. Infrared Microtechniques

The principles of microscopy can be used to pass IR radiation
through microsamples such as single fibers. Valuable information
can be obtained in this way which would be difficult to get in any
other manner. An IR microscope has been designed and constructed
for this purpose as an attachment to a single-beam commercial IR
spectrometer. Coates and co-workers [12] obtained good IR absorption
spectra of a single acrylic and nylon fiber by using this attachment.
This procedure has been reviewed by Hurtubise [13], Elliott [14],
and Zbinden [15].

G. Thin Fiber Films

Zhbankov and Ermolenko [16] described the development of a
technique whereby they have prepared thin fiber films under high
pressure without the use of alkali halide. Zhbankov later reported
on this technique in a book published by Consultants Bureau [17].

Knight and co-workers at the George Institute of Technology
[18] reported a similar method of sample preparation to obtain IR
spectra of short staple fibers. A thin film of parallel fibers is
obtained under 2 to 12 tons pressure. Adapters are made to fit into
a commercial KBr pellet press to hold the fibers while pressure is

applied for 1 to 5 min. It is reported that spectra of fibers obtained by this method are as good as, or better than, those obtained
by the KBr disk or FMIR techniques.

Tirpak and Sibilia [19] reported a new technique of sample preparation which involves winding a single filament between two salt
windows. The windows are separated by a spacer which is the same or
slightly greater in thickness than the fibers.

III. APPLICATIONS OF INFRARED SPECTROSCOPY

Within the past 20 years, the application of IR spectroscopy to textiles has increased many fold. This is evident from the great increase in the number of articles published on the subject.

Analysis by IR spectroscopy is important because an IR spectrum
is as characteristic of a compound as a fingerprint is to a human
being. There are a number of areas in which IR analysis has been
applied to textiles.

A. Identification of Fibers

Fibers, fabrics, and yarns can be identified quite readily after
the sample is prepared by one of the methods described above. The
sample preparation method used depends on the physical characteristics of the fiber and the equipment one has available. After an IR
spectrum has been obtained of an unknown, the next step is to attempt
to match the unknown spectrum with a standard spectrum. A number of
good reference spectra sources is listed in the bibliography for
this chapter [11,20-23].

The identification of fiber blends by IR spectroscopy has been
reported by Wharton and Forziati [24].

B. Identification of Finishes, Plasticizers, Additives, and Coatings

A method for the analysis of film-forming finishes has been
presented as a possible test method for the textile and chemical
industry consideration [25]. The finishes are separated and the

individual IR spectra of the different finishes are compared with
standard spectra.

The AATCC Test Method 94-1969 includes a confirmatory test for
different materials used as sizes [26]. The method includes a recommended sample preparation procedure and IR absorption spectra of a
number of sizes.

Forziati et al. reported a method for the identification of
textile coatings by IR spectroscopy [27].

Morris et al. reported the development of a method for identifying finishes [28]. After successive extractions, the individual
fractions have been analyzed to determine the finishes present. The
report includes methods for analyzing for both fluorochemical and
flame-retardant finishes.

Nettles demonstrated the utility of IR spectroscopy in the identification of surfactants [29]. The classes of surfactants covered
are fatty acid soap, sulfuric acid esters, and ethylene oxide derivatives, as well as blends of surfactants and demulsified solvents.

Hummel [30] has published a very valuable collection of IR absorption and chemical methods for the analysis of surface-active
agents.

Carlsson et al. reported the use of FMIR IR spectroscopy in the
identifications of coatings on fabrics, surface finishes on textile
products, and the photo-oxidation products of polypropylene [31].

C. Studies of Chemical
     Modification of Cellulose

O'Connor et al. used the KBr technique to obtain IR spectra of
cellulose that have been chemically modified [32]. They studied
the changes which occurred in hydrogen bonding, crystallinity, and
cross-linking during chemical modifications, including oxidation,
esterification, and etherification. They tabulated 50 IR absorption
bands that are of particular significance in the study of cellulose
chemistry. Others have used IR spectroscopy to study the chemical
modification of cotton [8,33,34]. Most samples were prepared for
IR examination by the KBr pressed disk method. McCall and co-workers

used FMIR spectroscopy to study cotton piece goods treated with
N-methylol-type reagent [35]. They emphasized the advantage of the
FMIR technique over the KBr pressed disk method in the examination
of surface coatings of fabrics. Simonian reported the IR spectra
of some cellulose reactants [36].

McCall et al. reported an analytical method for identification
of nitrogenous cross-linking reagents on cotton by IR spectroscopy
[37]. They used four different techniques: the conventional KBr
pressed disk, a differential disk, acid hydrolysis, and FMIR.

The effect of sodium hydroxide on cotton was studied. The IR
spectra showed that there is a change in the hydroxyl band of the
cotton. Apparently, some hydrogen bonds were broken [11,17,38].

D. Chemical Modification
of Other Textiles

Stein and Guarnaccio studied the oxidation of hair, callus,
and wool by IR spectroscopy. They used the KBr pressed disk method
of sample preparation [39].

Studies of the application of IR spectroscopy to wool were reported in a two-part article in *Wool Science Review* [40]. In the
second part of the article, the spectrum of β-keratin was shown and
interpreted. Hydration and deuteration of β-keratin were studied.

Cassels reported an interesting technique for examining the IR
spectra of the pyrolysis products of 15 different synthetic fibers
[41]. The spectra of the pyrolyzed fibers are compared with FMIR
spectra of the fibers themselves. The Wilks Model 40 Pyro-Chem accessory was used to pyrolyze the samples prior to IR examination.

E. The Interpretation
of Textile Spectra

As described above, the matching of an IR spectrum of an unknown
to the spectrum of a known fiber is a straightforward technique. Use
of standard collections of spectra make this operation possible.
When the unknown and known spectra are superimposable, a positive
identification can be made.

Other methods of identification from IR spectrum of an unknown require skill and experience on the part of the analyst, whether he is studying a fiber or a finishing material. The absorption bands of the specific organic functional groups can be studied and a structure implied in this way. To use this method, the analyst must be quite familiar with the chemistry of the finishes or fibers involved before a positive identification can be made.

O'Connor tabulated IR band frequencies that are unique to certain synthetic and natural fibers [11]. Zbinden gives a survey of references to IR spectra of individual polymers [15]. In most cases, he breaks down the references into (1) general discussions of spectra, (2) discussions of specific bands, (3) polarization measurements, (4) crystallinity effects and measurements, (5) analytical measurements.

In the bibliography are listed additional sources which this author has found to be valuable in helping to elucidate the IR spectra of unknowns [42-45].

F. Identification of Textile Dyes

Abrahams and Edelstein applied the IR spectroscopic technique to the identification of an ancient dyed textile [46]. The dyed materials came from Bar Kochba finds in the Judean Desert, and were possibly used by the Bar Kochba rebels as clothing in 135 A.D.

The application of IR spectroscopy to dye house problems has been studied. Contaminants such as oils, fats, waxes, and greases on cotton textiles can cause spotty or uneven dyeing. Rayburn and co-workers proposed a method for the isolation and identification of such contaminants by IR spectroscopy [47].

IV. CONCLUSION

A search of the textile journals shows a clear trend to the continuing increase in the application of IR spectroscopy to textiles. The same cannot be said for Raman spectroscopy. As yet, there is not a single reference to the application of Raman spectroscopy to textiles. Only time will tell whether Raman spectroscopy will follow the path made by IR spectroscopy since the 1950s. It is evident that IR spectroscopy has become an important method for the textile chemist, as an analytical tool in fiber and finish identification and as a research tool in the creation of new fibers and finishes. Both ASTM Committee D-13 and the American Association of Textile Chemists and Colorists have written test methods employing IR spectroscopy.

I cannot finish this review without mentioning the following references which belong in the library of any textile chemist who works with IR spectroscopy [11,17,48,49].

REFERENCES

1. J. W. Rowen, C. M. Hunt, and E. K. Plyler, *J. Res. Natl. Bur. Std.*, *39*, 133 (1947).

2. F. H. Forziati, W. K. Stone, J. W. Rowen, and W. D. Appel, *J. Res. Natl. Bur. Std.*, *45*, 109 (1950).

3. R. T. Conley, *Infrared Spectroscopy* (2nd ed.), Allyn and Bacon, Boston, 1972, p. 50.

4. D. N. Kendall in *Sample Preparation Procedures, Applied Infrared Spectroscopy* (D. N. Kendall, ed.), Reinhold, New York, 1966, p. 141.

5. U. Schiedt and H. Remwein, *Z. Naturforsch, 7b*, 270 (1952).

6. M. M. Stimson and M. J. O'Donnell, *J. Am. Chem. Soc., 74*, 1805 (1952).

7. M. R. Harvey, J. E. Stewart, and B. G. Achhammer, *J. Res. Natl. Bur. Std., 56*, 225 (1956).

8. R. T. O'Connor, E. F. DuPré, and E. R. McCall, *Anal. Chem., 29*, 998 (1957).

9. F. H. Forziati and J. W. Rowen, *J. Res. Natl. Bur. Std., 46*, 38 (1951).

10. J. Fahrenfort, *Spectrochim. Acta, 17*, 698 (1961).

11. R. T. O'Connor in *Absorption Spectroscopy, Analytical Methods for Textile Laboratory* (J. W. Weaver, ed.), 2nd ed., American Assoc. Textile Chemists and Colorists, Research Triangle Park, North Carolina 1968, p. 312.

12. V. J. Coates, A. Offner, and E. H. Sigler, Jr., *J. Opt. Soc. Am., 43*, 984 (1953).

13. F. Hurtubise, *Can. Textile J., 76*, 53 (1959).

14. A. Elliott in *The Infrared Spectra of Polymers, Advances in Spectroscopy,* Vol. 1 (H. S. Thompson, ed.), Wiley (Interscience), New York, 1959, p. 218.

15. R. Zbinden, *Infrared Spectroscopy of High Polymers,* Academic Press, New York, 1964, p. 166.

16. R. G. Zhbankov and I. N. Ermolenko, *Izv. Akad Nauk BSSR, Ser. FizTekhn, 1,* 15 (1956) (in Russian).

17. R. G. Zhbankov, *Infrared Spectra of Cellulose,* Consultants Bureau, New York, 1968, p. 23.

18. J. A. Knight, M. P. Smoak, R. A. Porter, and W. E. Kirkland, *Textile Res. J., 37,* 924 (1967).

19. G. A. Tirpak and J. P. Sibilia, *J. Appl. Polymer Sci., 17,* 643 (1973).

20. R. D. Morrison, Report of AATCC Committee RR24 Fiber Analysis Test Methods. *Am. Dyestuff Rep., 52,* 43 (1963).

21. R. A. Wilks, Jr., and I. R. Mayhem, Paper presented at the 15th Mid-American Spectroscopy Symposium, Chicago, Ill., June 2-5, 1964.

22. *A.S.T.M. Standards on Textile Materials,* 33rd ed., American Society for Testing and Materials, Philadelphia, 1962, pp. 135-141.

23. Sadtler Research Laboratories, Philadelphia, Pa.

24. M. K. Wharton and F. H. Forziati, *Am. Dyestuff Rep., 50,* 515 (1961).

25. B. Norwick, *Am. Dyestuff Rep., 50,* 329 (1961).

26. AATCC Test Method 94-1969, Finishes in Textiles: Identification. *Technical Manual of the Amer. Assoc. of Textile Chemists & Colorists,* Vol. 48, American Association of Textile Chemists and Colorists, Research Triangle Park, North Carolina 1972, pp. 64-68.

27. F. H. Forziati, R. T. Hite, and M. K. Wharton, *Am. Dyestuff Rep., 49,* 103 (1960).

28. N. M. Morris, E. R. McCall, and V. W. Tripp, *Tex. Chem. and Color., 4,* 283 (1972).

29. J. E. Nettles, *Tex. Chem. and Color., 1,* 430 (1969).

30. D. Hummel, *Identification and Analysis of Surface-Active Agents by Infrared and Chemical Methods,* Vols. 1 and 2, Wiley (Interscience) New York, 1962.

31. D. J. Carlsson, T. Suprunchuk, and D. M. Wiles, *Can. Textile J., 87,* 73 (1970).

32. R. T. O'Connor, E. F. DuPré, and D. Mitcham, *Textile Res. J., 28,* 382-392, 542-554 (1958).

33. R. T. O'Connor, E. R. McCall, and D. Mitcham, *Am. Dyestuff Rep., 49,* 214 (1960).

34. V. W. Tripp, E. R. McCall, and R. T. O'Connor, *Am. Dyestuff Rep., 52,* 598 (1963).

35. E. R. McCall, S. H. Miles, and R. T. O'Connor, *Am. Dyestuff Rep., 56,* 400 (1966).

36. J. Simonian, *Am. Dyestuff Rep., 53,* 248 (1964).

37. E. R. McCall, S. H. Miles, and R. T. O'Connor, *Am. Dyestuff Rep., 57,* 35 (1967).

38. G. Celikiz, M.S. thesis, Georgia Institute of Technology, 1956.

39. H. H. Stein and J. Guarnaccio, *Textile Res. J., 29,* 492 (1959)

40. G. L. Stott and E. G. Bendit, *Wool Science Review,* Research Dept., International Wool Secretariat, London, Jan. 1969, pp. 35-48.

41. J. W. Cassels, *Appl. Spectry., 22,* 477 (1968).

42. R. E. Kagarise and L. A. Weinberger, *Infrared Spectra of Plastics and Resins,* NRL Report 4369, Naval Research Laboratory, Washington, D. C., 1954.

43. R. A. Nyquist, *Infrared Spectra of Plastics and Resins* (2nd ed.), The Dow Chemical Company, Midland, Michigan, 1961.

44. H. Hausdorff, *Analysis of Polymers by Infrared Spectroscopy,* The Perkin Elmer Corporation, Norwalk, Connecticut, 1951.

45. J. Haslam and H. A. Willis, *Identification and Analysis of Plastics,* D. Van Nostrand, Princeton, New Jersey, 1965.

46. D. H. Abrahams and S. M. Edelstein, *Am. Dyestuff Reporter, 53,* 19 (1964).

47. J. A. Rayburn, J. M. Fields, N. B. Moore, and J. W. Weaver, *Am. Dyestuff Reporter, 52,* 295 (1963) and *53,* 19 (1964).

48. R. T. O'Connor, *Instrumental Analysis of Cotton Cellulose and Modified Cotton Cellulose,* Marcel Dekker, Inc., New York, 1972.

49. R. T. O'Connor, *Infrared Absorption Spectroscopy of Cellulose and Cellulose Derivatives in Developments in Applied Spectroscopy,* Vol. 5 (L. R. Pearson and E. L. Grove, eds.), Plenum Press, New York, 1966, pp. 129-156.

Date Due
